Fossils
The Key to the Past

THIRD EDITION

Richard Fortey

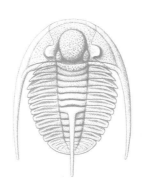

Smithsonian Institution Press, Washington, D.C.
in association with The Natural History Museum, London

For picture credits please see page 4, all other images are copyright of
The Natural History Museum, London and taken by the Museum's Photo Unit.

For copies of these and other images please contact The Picture Library,
The Natural History Museum, Cromwell Road, London SW7 5BD
View the Picture Library website at www.nhm.ac.uk/piclib

Published in the United States of America
by the Smithsonian Institution Press
in association with The Natural History Museum, London
Cromwell Road
London SW7 5BD
United Kingdom

Library of Congress Cataloging-in-Publication Data
Fortey, Richard A.
    Fossils : the key to the past / Richard Fortey.—3rd ed.
        p.   cm.
    Includes bibliographical references and index.
    ISBN 1-58834-023-6 (alk. paper) — ISBN 1-58834-048-1 (pbk. : alk. paper)
        1. Fossils. 2. Paleontology. I. Title.
    QE711.3.F67   2002
    560—dc21   2001049439

Manufactured in Singapore, not at government expense
09 08 07 06 05 04 03 02   5 4 3 2 1

Edited by Rebecca Harman
Designed by Stewart Larking
Reproduction and printing by Craft Print, Singapore

Front cover main image: Fossils sea-bed assemblage with corals, bryozoans,
brachiopods and trilobites. Silurian, Wenlock Limestone, Dudley, West Midlands, UK.

Contents

· · · · · · · · · · · ·

Preface   5

Chapter 1: Buried in the rocks   7

Chapter 2: Setting the stage: time and change   27

Chapter 3: Rocks and fossils   67

Chapter 4: Bringing fossils back to life   93

Chapter 5: Origin of life and its early history   117

Chapter 6: Evolution and extinction   143

Chapter 7: Fossils in the service of humans   187

Chapter 8: Making a collection   209

Further Information   225

Glossary   226

Index   228

Acknowledgements

Special thanks to Chris Stringer for helping me update the hominid section, and to Brian Rosen for supplying photographs from his personal collection. Thanks also to my colleagues who looked out new pictures, and who checked caption text. Thanks, too, to the other Richard for being a very efficient bookies runner.

Picture credits:

p.12 © H Whittington; p.13 © W Stürmer; p.16 © Angela Milner; pp.20 & 43 © Alan Timms; p.46 Illustrated Image/© NHM; p.66 © Brian Rosen; p.76 © John Chapman, p.77 Illustrated Image/© NHM; p.81 left Ray Burrows/© NHM; p.81 right © Peter Skelton; p.83 © Linda Pitkin; p.90 Illustrated Image/© NHM; p.92 The National Geological Museum of China/ © NHM; pp.96 & 97 Ray Burrows/© NHM; p.100 © Andrew Milner; p.103 The National Geological Museum of China/ © NHM; pp.104, 105, 108, 111 Ray Burrows/© NHM; p.114 top © Lars Ramsköld; p.114 bottom Ray Burrows/© NHM; p.116 Courtesy of NASA; p.119 © Don Cowan; p.121 Courtesy of NASA/JPL/Caltech; p.123 © Bill Schopf; p.126 © Pamela Reid; p.125 top © Brian Rosen; p.129 © N Butterfield; pp.132, 135, 138 bottom, 142, 148, 149 & 184 Ray Burrows/© NHM; p.190 © Mark Purnell; p.201 Ray Burrows/© NHM; p.202 © Alan Timms

Preface

· · · · · · · · · ·

I wrote the original edition of *Fossils: The Key to the Past* nearly 20 years ago. Since then there has been many new discoveries in palaeontology which I have attempted to incorporate in subsequent versions. This is true also of this latest edition for which the illustrations have been revisited, and the original structure slightly amended.

Palaeontologists continually review their vision of the past, and it could well be that even in three or four years time there will be further discoveries leading to new understanding. Like all sciences palaeontology moves forward or it dies. Nonetheless there is still a place for a straightforward introduction to the study and meaning of fossils: how they help us understand the geological past, how their fascination is always more than just 'stamp collecting'. Even though new discoveries continually add to our comprehension, the fundamentals of science do not change so radically.

The purpose of the book remains the same: to stimulate the reader to further study and enjoyment of our geological legacy.

Richard Fortey
January 2002

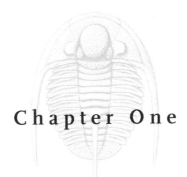

Chapter One

BURIED IN THE ROCKS
. . . .

FOSSILS are the remains of prehistoric animals or plants. They used to be regarded as curious freaks of nature and explanations of the way they were preserved could be fanciful. Ammonites found at Robin Hood's Bay on the Yorkshire coast, UK were supposed to have been snakes turned to stone by St Hilda. Enterprising local craftsmen embellished the fossils by carving on an appropriate snake's head. As it became clear that fossils were present in the rocks as a result of natural processes, the mechanism of fossilization became better understood. The majority of fossils comprise the hard parts of the animal or plant, or structures resistant to decay. Shells, bones, wood and teeth are all likely to be preserved as fossils. Molluscs and mammals are thus likely to have a good fossil record, worms and amoebae a poor one. The clusters of dead shells one finds in rock pools on the beach — limpets, crab claws, a winkle or two, a broken sea urchin — are typical of the sort of debris that may become fossilized. Many fossils result from the cast off, outgrown or broken shells of marine animals. Such shell material

above: The ammonite
Dactylioceras, from Jurassic rocks
(Yorkshire, UK) with
a snake's head added by
a sculptor.
opposite: Excavation of the
dinosaur *Baryonyx walkeri*, at
Ockley brick pit (Surrey, UK)
in 1983.

opposite: The Silurian graptolite *Dictyonema retiforme* (425 Ma old). *Dictyonema* formed a cone-shaped colony in life, and is flattened from above (top), and from the side (bottom). The top specimen is from Hamilton (Ontario, Canada) and the lower specimen is from an area between Niagara Falls and Lewiston (New York, USA).

becomes covered with sediment and can be said to be a fossil from that stage on. Many fossils more than 1 million years old can look remarkably like shells picked up from the beach today. Most shells are porous and frequently the small pores in the fossilized material become the site for deposition of minerals, which make the fossil more dense than the original shell of the animal.

TURNED TO STONE

As sediments pile up at the site of deposition their coherence increases, some of the water they contain is expelled, and many subtle chemical changes occur to produce rock types such as shale, sandstone and limestone. In the harder rocks, such as limestone, fossils buried with them retain much of their original shape and convexity, but shales frequently become compressed, and as this happens the fossils become flattened. Sometimes the same fossil species can have a very different appearance according to whether it has been preserved in hard rock or flattened. Further changes can happen to fossils as a result of their sojourn in the rocks. Most shells are made of calcium carbonate, which is soluble in carbonated water. In porous rocks, especially sandstones through which groundwater travels, the shell can be dissolved away. The fossil is not lost in this case because the rock itself will have taken an impression of the shell, just as a fingerprint can be impressed into clay. When such a rock is split open the inside of the shell will be preserved as an internal mould, while the other half (the counterpart) of the specimen will contain the impression of the exterior details. To obtain a reconstruction of the whole shell you need both halves of the fossil specimen. The golden rule of

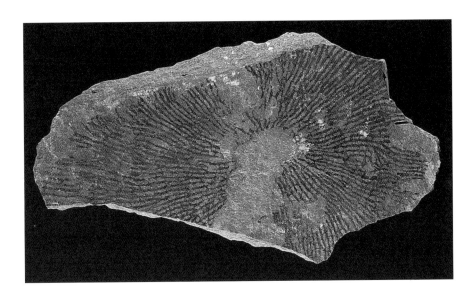

right: Internal (far left) and external (right) moulds of the Ordovician trilobite *Placoparia* from the Czech Republic.

collecting is: Never throw away the counterpart. Many a vital specimen has gone tumbling down a scree slope when this rule has been ignored. In some cases the cavity left after the shell has been dissolved is replaced by another mineral; if you are very lucky the mineral might be opal.

Sometimes these secondary replacements can be advantageous to the palaeontologist. Fossils with shells originally of calcium carbonate may be replaced by the mineral silica. If this happens in a limestone, the enclosing rock can be dissolved in acid and, because the silica is insoluble, the replaced fossil will remain. In this way remarkably delicate details of the original fossils can be preserved such as spines and other features, which are impossible to dig out mechanically. A rarer type of preservation is perfect replication almost molecule by molecule of the original fossil, so that minute details of the microstructure are determinable. Such petrifaction is usually in fine-grained silica. One famous example is the Rhynie Chert, a Devonian petrifaction of some of the earliest land plants. The perfection of preservation is such that sections of these early plants clearly show the individual cells in the plant tissues. Such specimens are of immense importance in revealing the most intimate details of the structure of extinct organisms, which can then be compared with some confidence with their living relatives for which there is complete information.

above: Silicified brachiopods showing the inside of the dorsal valve in the upper picture, and the outside of the ventral valve in the lower, with long, delicate spines which would normally break off in the rock.

GEOLOGICAL MIRACLES

In very rare cases it is not only the hard parts of animals and plants that are preserved in the fossil record. These geological 'miracles' also preserve the impressions of the soft organs of animals that would normally decay without trace. There are always special geological circumstances in such cases. One of the most famous, and oldest, of these is the Burgess Shale from British Columbia, Canada. This is a black, fine-grained Cambrian shale about 515 million years old, with a host of wonderfully preserved fossils, many of which are unknown anywhere else in the world. It affords us a unique glimpse of almost the whole spectrum of life at this very early stage in its history, and shows many animals not preserved under normal circumstances in the fossil record. Trilobites and other arthropods are preserved with all their limbs, antennae, and even their gut contents. Some of the animals are hard to match with any living organisms, and may represent kinds of creatures that have long since vanished from the Earth. Many kinds of worms are present in the fauna of the Burgess Shale. The soft parts have been preserved as the thinnest of films, and the worms were probably buried rapidly, before the soft parts could decay, in an environment where they could not be shredded to pieces by scavenging organisms. A similar occurrence in the Devonian Hunsrück Shale, Germany, has the soft parts of animals preserved as a thin film of the mineral iron pyrites. In this case the structure of the soft parts can be studied using X-rays, which pick out the iron pyrites; the structures can be photographed in the rock even though they are not visible on the surface.

left: The Cambrian trilobite *Olenoides*, from the Burgess Shale (western Canada), showing the traces of its limbs and other soft parts not normally preserved in the fossil state.
right: *Marrella*, also from the Burgess Shale, a peculiar primitive arthropod unknown elsewhere.

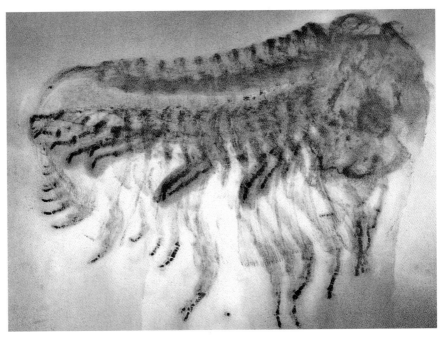

above: A phacopid trilobite from the Devonian Hunsrück Shale (Germany). Fine details of the limbs are preserved.

The early bird *Archaeopteryx* is preserved in a fine-grained limestone, creamy brown in colour, which was used for the manufacture of lithographic blocks, found near the Bavarian town of Solnhofen. This could scarcely be more different in appearance from the Burgess Shale, but like it the lithographic limestone retains spectacular remains of a whole host of animals with scarcely any fossil record elsewhere. The fossils are Jurassic in age, and many are related, distantly, to animals still living. Besides *Archaeopteryx* there are reptiles, dragonflies, relatives of the horseshoe crab *Limulus*, sea spiders and mammals, in effect a mixture of terrestrial and marine life. The Solnhofen deposits are believed to have accumulated on the boundary between land and sea, probably in a lagoon, where a sticky, limy mud was accumulating. Flats of this mud were probably exposed at low tide, and at this time *Archaeopteryx* became entrapped. The fine mud was ideally suited to entomb the remains of the bird and also to take an impression from delicate feathers that would otherwise have decayed without trace. It is fortunate that this happened, because otherwise there would be doubt as to whether or not *Archaeopteryx* was really a bird. Some scientists claim that the *Archaeopteryx* specimens are fakes, manufactured by impressing feathers of living birds on a prepared surface surrounding genuine dinosaur fossils. This claim is inherently improbable given the discovery of fossils of *Archaeopteryx* from time to time over many decades (it would imply a scientific conspiracy of massive proportions, and without commensurate motivation). Furthermore, careful examination of the fossils has not revealed any evidence of such fakery.

Rare, geological miracles like the Burgess Shale and the Solenhofen limestones are of exceptional palaeontological importance, the information they yield is like a floodlight on the past, whereas most geological sites are more like an intermittent flashlight. The geologically youngest of such exceptional fossils are the frozen mammoths of Siberia, dating from late in the last Ice Age. These extinct giants were apparently frozen so quickly that their hair and flesh are preserved almost as it they had been frozen for the supermarket. Speculation about regenerating one of these remarkable animals from their frozen cells is probably over-optimistic.

Amber ornaments were extremely popular in Victorian times, and the most prized of these had small insects displayed within the amber droplets. Apart from its beauty, amber preservation is another exceptional occurrence where animals not normally preserved as fossils are found in abundance. Amber is generally 70 million years old or less (some is as old as 120 million years). The majority of the insects encased within it are related to living forms and are of great importance in understanding the genesis of the most diverse group of living invertebrates. Amber is

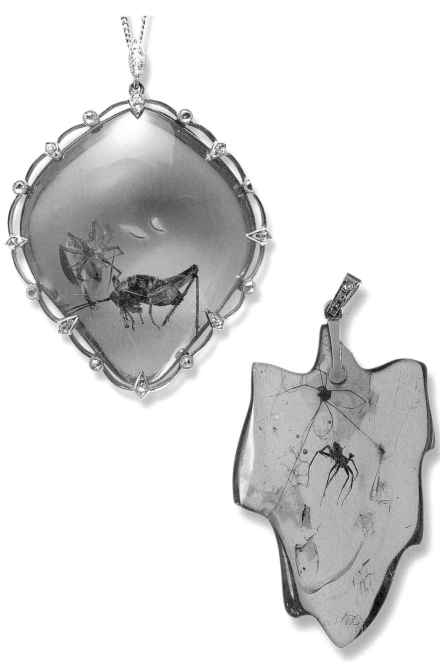

above: Baltic amber pendants, containing a spider and a cri[...]
spider and a harvestman (bottom).

opposite: This Middle Jurassic 147 m (482 ft) long sauropod trackway (175–165 Ma old), from a site in Portugal, is the longest trackway on record. The track was made by an animal walking in a straight line across a muddy estuary.

produced as a resin oozing from the branches and trunks of coniferous trees. Insects and spiders became trapped in this resin, and enclosed within it as more resin was added and solidified. Hardened resin is extremely tough, and so has a high chance of being fossilized, carrying with it its cargo of preserved insects. Lumps of amber eventually found their way into sediments, from which they can be recovered like any other fossil. The same process of entrapment still goes on today, and the Recent (geologically speaking) hardened resin (copal) can be sold as ersatz amber.

TRACE FOSSILS

A fascinating branch of palaeontology is concerned with the traces left by the activity of extinct animals — trace fossils (ichnofossils). These can be footprints, like the tracks of dinosaurs, left by animals as they grazed the sediment looking for food, or burrows made by animals escaping from a predator or laying their eggs. In many cases the animal itself is not preserved, and although it may be possible to deduce what it was doing, it is often not possible to say what animal was responsible for making the track. Huge, three-toed tracks of dinosaurs can tell us a lot about the gait of the animal that made them, or for example what its stride was, or whether the front limbs touched the ground. Many tracks are inconspicuous; labyrinthine or braided paths made by worms, and some of these worm 'diggings' are the only fossil record we have of these creatures.

Trace fossils are abundant even in the early Cambrian. They are especially numerous in sandy rocks, which otherwise lack body fossils. One of the Cambrian occurrences is the famous 'Pipe Rock' of the north west Highlands, UK: the pipes that give the rock its name are closely packed straight tubes which were made by some sort of worm. Other beds of rock in the same formation contain U-shaped or funnel-shaped tubes. The tracks made by trilobites are numerous in rocks of Cambrian and Ordovician age. These include winding trails and short burrows, some of which contain the impressions left by the legs of the trilobites. Trace fossils are given names, just like body fossils; the trilobite tracks are called *Rusophycus* or *Cruziana*.

An explanation is needed about the preservation of tracks: most tracks are dug into a sediment surface, an overlying layer of sediment then fills up the

above: Upper Cambrian trilobite tracks named as *Cruziana semiplicata*, thought to be made by a trilobite called *Maladioidella*, from Oman.

above: Pliocene pebble with concentric iron-staining broken to reveal borings, one with the bivalve *Hiatella*, which produced the boring, inside.

tracks, so that the cast formed by this infilling is a positive impression of the track itself. Burrows are dug deep into the sediment, and they often fill, once vacated, with more sediment of a different colour. Collecting past tracks can be almost as instructive as collecting the fossils of shells. Different kinds of animals live in different environments, and leave differing evidence of their activities. By studying tracks it is possible to find out a lot about the animals that lived on the former sea floor, even without remains of the animals themselves. Tracks should be distinguished from borings. These are made into a hard medium, such as rocky surfaces or wood. Some bivalved molluscs specialize in boring into hard rocky surfaces, and these too can be found as fossils, usually sitting at the end of their self-made caves.

above: Folded rocks resulting from tectonic processes may contain fossils that have been distorted from their true shape.

HISTORY OF FOSSILS WITHIN THE ROCK

Once they are incarcerated in the rocks, fossils are passive passengers, and what subsequently happens to the rocks also affects the fossils themselves. Not all rocks lie undisturbed until the hammer cracks open the booty they contain. Initially the fossils usually lie parallel to the bedding planes — the surfaces parallel to the surface of deposition. Some fossils remain in this attitude as the rocks are uplifted above the former ocean bed, to become exposed in cliffs, quarries or cuttings. More usually the uplift process, which may be connected with Earth movements (see Chapter 2), also produces tipping and gentle folding of the rocks, so that the bedding planes are no longer horizontal. This does not affect the fossils, which can be collected in the usual way. Where the Earth movements are more violent, the rocks may be squeezed and distorted, and so are the fossils contained inside. This is particularly the case in soft rocks, like shales. Older fossils are more likely to be found in this condition, because they have had more time to be involved in violent events, but many have escaped such action, and are as well preserved now as they were when they were first fossilized. Distortion includes stretching and twisting, and this results in a lot of the finer details of fossils being destroyed. Palaeontologists would like to have perfectly preserved material to work from, but often distorted fragments are all that is available from a large area, and they have to be identified as best they can.

The process of distortion does not stop with stretching and bending. As the rocks are squeezed progressively they sometimes become heated strongly and under these conditions the rock itself begins to change. This can have the effect of removing the fossils completely, the small ones first, then the most robust. Even so, fossils have been known to survive the most intense heating and high pressure. A very common secondary change that occurs in shaly rocks under pressure is that they develop a cleavage. This means that they start to split, not along the bedding planes on which the fossils lie, but at a high angle to the bedding, sometimes at right-angles. The kind of black or purple slates used for roofing are cleaved rocks of this kind. You can often see stripes passing across such slates and these are sections through the original bedding. It is no use splitting open such rocks in the hope of finding fossils parallel to the cleavage, although you might see a section through one if you are lucky. Fossils can be recovered with much hard work by smashing the slates in such a way that part of the original bedding is exposed. They are usually in a rather sorry state by the time they are found.

BITS AND PIECES

Many fossils are only fragments of the whole animal or plant. To piece together the whole organism is rather like doing a jigsaw puzzle without the benefit of the complete picture to work towards. Piece has to be added

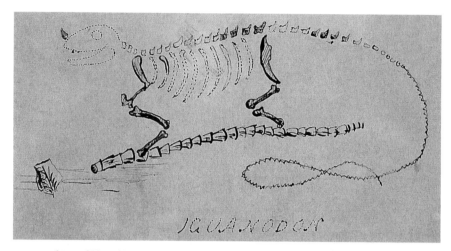

above: Gideon Mantell's original restoration of *Iguanodon*. Based on the skeletal anatomy of a modern lizard, Mantell placed the spike-like thumb of *Iguanodon* on its nose, like the horn of a rhinoceros. This misconception persisted for many years.

to piece, and the larger and more fragmentary the animal, the more the result will be in question. Not surprisingly mistakes have been made. The first reconstruction of the dinosaur *Iguanodon* finished up with its thumb on its nose! Trilobites are much more commonly found as pieces rather than as whole animals, and many kinds are known only from heads or tails until one day a lucky collector turns up a whole one. The problem is particularly acute for large plants, because there is nothing very obvious to connect the root with the trunk, or the trunk with the leaves if they are preserved in different places (and they usually are). Flowering structures, not being easily fossilizable, are sometimes even more difficult to assign to the plant. The result is that different names are given to different organs, one for the root, one for the bark and so on. Eventually, when the links are made, one name suffices for the whole plant.

There are many problems with piecing together fossils: distorted fossils have to be restored to their original shape, left valves of clams have to be matched with right, vertebrae and limb bones need to be placed together in the correct order. Many fossils are only known from fragmentary material, and it may be years (if ever) before the next piece is discovered. Just as there are rare and abundant species today, so there are rare and common species of fossils. Some species are so abundant that one is almost certain to turn up a specimen if rocks of the right age are hammered, others are so rare that a collector may go to a quarry one day and find a specimen, but visit the same place for years afterwards without finding a second one. This is all part of the particular fascination of palaeontology; you never know quite what will turn up. The discovery of new kinds of fossils is a regular

above: A team cleaning the exposed elements of the fore and hind limbs of a sauropod dinosaur at an excavation site in Niger, 1988–9.

occurrence, even in rocks that have been investigated by many collectors before. Each new discovery patches in a little of the story of life that was previously obscure. Even now many parts of the world are being explored for the first time, and the new palaeontological finds show little sign of abating. Professional palaeontologists often feel swamped by the sheer variety of different fossils there are to deal with.

Although they are fascinating just as attractive objects, fossils are of enormous practical importance in interpreting the history of the planet as a whole. The study of fossils links in with other branches of geology and biology; indeed a knowledge of related sciences is essential to appreciate the significance of fossil finds. Recently, for example, the study of 'chemical fossils' has

above: Sir Charles Lyell, 1797–1875, who believed that the Earth had been subject to gradual change through the action of processes such as earthquakes and erosion.

become an important new field. Such chemical fossils are often the remains of decayed organic material, which provide enduring organic molecules left in the rock, and these can tell us what kind of organism made them. Study of the variations in the isotopes of carbon and oxygen can elucidate past atmospheric and oceanic conditions. Sophisticated equipment capable of measuring minute quantities of organic material is becoming more and more important to the professional palaeontologist. The next two chapters will explore some of the connections with other geological sciences, to show how the animal and plant life of the Earth is intimately bound up with the story of the rocks themselves, and the configuration of the continents and oceans. As Lyell recognized, the forces of physics have been the same through geological time, so the processes that formed the present Earth are not beyond our understanding. But the unchanging physical laws operate on a thoroughly mutable world, and the configuration of land and sea has changed repeatedly. For hundreds of millions of years living organisms have altered in harmony with the world, and in the process have themselves transformed it.

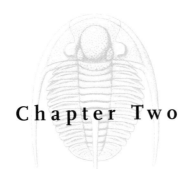

Chapter Two

SETTING THE STAGE: TIME AND CHANGE

THE EARTH is approximately 4500 million years old. This figure is now generally accepted, but has only become so in the last few decades, especially since the age of the Moon was determined. If we return to the 19th century, at the time when Hutton saw that the Earth must be 'immeasurably' old for present processes to account for all the varied features of the rocks and their immense thickness, the business of putting an actual age on the Earth was very problematic. Once the idea of Creation within a few thousand years disappeared, it was obvious that the Earth had to be much older. But how much? Millions of years certainly, but 10 million, 100 million, or 1000 million? The fact is that it is impossible for the human mind to grasp such lengths of time as these. It was easy enough to realize that a long span of time was needed, but difficult to devise methods of assessing the actual amount of time involved.

opposite:
Megatherium, the giant sloth, (about 6 m or 20 ft from head to tail) endemic to S America when the continent was geologically isolated. Some species of giant sloth moved into North America.

GEOLOGICAL TIME

The history of the Earth is written in the rocks, and the rocks can be divided into natural units. Once the sequence of rocks and fossils contained within them was worked out, it was possible to say something about the different kinds of organisms that had populated the Earth and replaced one another over geological time. In the 19th century, fossils were found in all rock units except the underlying Precambrian rocks, which seemed to be barren of all fossil life. How much of the Earth's history did these represent? At least one school of thought believed that these ancient rocks were partly the original crust of the Earth (many of them being crystalline) and an approximation of the age of the Earth could be obtained by guessing the length of time taken for the deposition of the Cambrian and later rocks. If the rocks accumulated at a fixed rate, then the maximum total thickness of the rocks should provide a basis upon which to measure the time it had taken to deposit them. At the time the subject was debated in various ways, in particular focussing on the suggestion that the rocks would be compressed by burial, and the resulting estimates of total deposition time ranged from less than 20 million years to more than 700 million years. The great physicist, Lord Kelvin, assumed that the Earth had cooled from a totally liquid state, and the time taken to produce the current condition of heat flow and solid crust would be between 20 and 40 million years. By the 1890s this seemed to be the most scientific answer.

At the same time the knowledge of the relative timescale of geology had come to resemble closely the scale we use today. The question of absolute (in millions of years before present) age is a different one, and the whole edifice of geology was built largely on relative relationships. The Silurian rocks were older than the Devonian rock, because they underlay them, and the fossil fauna could be shown to be different; the Devonian in turn underlay the coal-bearing Carboniferous, and these rocks underlay the Permian formations, and so on up the geological column.

The fossil faunas also seemed to change in a radical way from one of these broad units to another, and what better way, therefore, to define the divisions of geological time? The development of the rock formations varied from one area to another, and in some cases provided the name of

above: An angular unconformity between two rock formations: Triassic rocks (horizontal) on Devonian rocks (inclined at about 40°), western England, UK.

the geological period, for example the Jurassic period was named after the Jura mountains. Some of the natural divisions of geological time were bounded by unconformities, angular breaks in the sequence of rocks. Rocks lying beneath an unconformity were uplifted and folded before the rocks lying on top of them were deposited. The divisions of geological time from the Cambrian onwards were established in Europe, and it is a measure of the acumen of the geologists who established them that the original concepts still survive today.

The 15 major divisions of Phanerozoic geological time (periods or systems) are shown on p.30. The first four of these (Cambrian, Ordovician, Silurian, Devonian) all take their names from British localities; Cambrian from the Latin word for Wales, Ordovician and Silurian from two of the old Welsh tribes, and Devonian from Devon. The Carboniferous rocks comprise the coal-bearing strata, and the Permian rocks are typically developed in the Perm mountains of Russia. Triassic rocks naturally fall into three (Tri-)

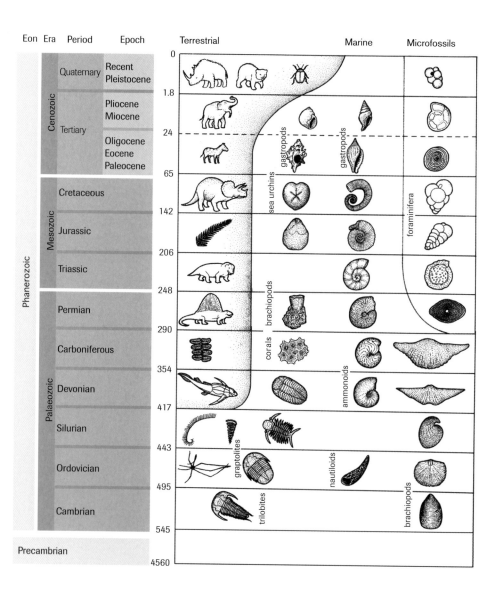

above: The main divisions of geological time (figures are in million years).
The geological timescale was originally established from the relative ages
of the rocks, partly based on fossils. Some of the important fossils for dating
are shown.

divisions in their characteristic development in Germany, and the Jurassic period was christened as mentioned above. The Cretaceous included the pure white limestone known as chalk (Latin: creta). For broader purposes it is still useful to talk about eras, three great divisions of fossil-bearing time: Palaeozoic (literally 'old life' encompassing the Cambrian to Permian periods), Mesozoic ('middle life' spanning the Triassic to Cretaceous) and Cenozoic taking us up to the present day.

The boundaries between the eras represent the most important extinction events the world has known, when much of the living world was replaced by new and different organisms, a graphic way of dividing the history of life into three portions. We can now add a fourth era, the Proterozoic (ancestral life) for the later part of the Precambrian, from which fossil organisms are now widely known. The Tertiary periods are shorter than the preceding ones, and were originally proposed by Lyell on the basis of how closely the fossil faunas resembled those of the present day. They include the Eocene ('dawn of the present'), the Miocene ('less than present') and so on. Finally the Pleistocene ('nearly present') includes the ice-formed deposits that have been draped like an irregular blanket over the earlier geology of the Northern Hemisphere, where the geological record merges into the recorded history of humans.

Once the language of geological time had been developed it was a great help in communicating the discoveries that had been made internationally, and in building up the first pictures of whole faunas for particular periods. Certain periods acquired popular descriptions, some of which seem to have stuck. The Devonian became the 'Age of Fishes', the Carboniferous the 'Age of Amphibians' and the Jurassic–Cretaceous the 'Age of Reptiles'. These tags have endured, and they do serve a purpose to emphasize some of the largest animals that lived in the respective periods, but they also serve to give a rather false impression of all the different biological events that were going on at the time. There is always a temptation to view the fossil record as if it were a kind of staircase with progressive steps leading upwards towards humans. This is misleading, because evolutionary activity has been unremitting in even the humblest of creatures, and dominance of the natural world by larger animals is only a matter of their conspicuousness.

The geological periods were soon to prove only the most general way of subdividing geological time. The periods themselves could be subdivided into segments which would enable a much more refined way of talking about the relative ages of organisms. For example, dinosaurs changed greatly during the Cretaceous period, and it was necessary to have a way of describing the timing of these changes. Periods were divided into early and late parts (mid-parts in many cases as well). The finest subdivision became (and remains) that of the zone. A zone is a division of geological time characterized by a particular assemblage of fossil species. It is a small segment of geological time through which the evolutionary history of various organisms pass. There will be a number of animals or plants that are unique to a specific zone, although others which have evolved more slowly may range through more than one zone. The name of the zone is taken from one of the most characteristic of its defining organisms. For example, in the Ordovician, graptolites are of importance in subdividing the rocks, and the 'Zone of *Nemagraptus gracilis*' is named after one species which is of widespread occurrence in its zone, although it is accompanied by other species characteristic of the same time period. The zone is another way of communicating the exact age (on the relative timescale) of a fossil, and, one hopes, means the same in all areas of the world. Some kinds of organisms have become more useful in the definition of zones than others, and the most useful have proved to be those which evolved rapidly (because they enable the time to be sliced finely) and which achieved wide geographic dispersal.

Much of the attention of geologists is devoted to trying to establish the age relationships between strata from widely different localities – the correlation of rocks, and this is the cornerstone of the branch of geology known as stratigraphy. It is not necessary to know the exact age in millions of years to be able to correlate, the basic statement is 'this rock is the same (or not the same) age as that one' whether the age of the rock in question is 2 million years old or 200 million years old. The fossil content acts as the clock.

The kinds of fossil organisms used as the basis for zones varies through geological time, as one group rises to prominence only to be replaced by another. It is the humblest of fossils that are often the most useful as the basis of the zonal schemes. In theory any organism with a good fossil record can be used for a zonal indicator, but obviously dinosaurs are far too large to be recovered from an average roadside exposure even though they may have evolved very quickly. Common fossils like ammonites, brachiopods and trilobites are used as zonal fossils, partly because they can be recovered from most (marine) sediments of the right age, and partly because they show enough variation through time to be readily recognized in the laboratory by the palaeontologist.

A whole separate branch of palaeontology has grown up around using the smallest of fossils as zonal indicators. This is known as micropalaeontology and is described further in Chapter 7. Small fossils are of particular use in dating rocks recovered from boreholes, where the narrowest of cores may yield large numbers of diagnostic fossils. Not surprisingly, this kind of palaeontology is much employed by oil companies and other commercial enterprises concerned with recovering mineral wealth from considerable depths. Dating the rocks, and correlating between boreholes, is a fundamental part of mineral and oil exploration. It is possible for several zonal schemes to exist side-by-side, referring to the same time period; one may be primarily concerned with microfossils, another with brachiopods or ammonites and so on. One system usually becomes the 'standard' for a particular geological period, often the one first proposed.

The most important fossils used as the basis of zones change through the geological column. In the Cambrian, trilobites are more widely used than any other group, largely because they were among the most varied and numerous of the invertebrates at the time. For the Ordovician and Silurian periods, graptolites have been widely employed for zonal purposes, although trilobites, brachiopods and other fossils are still used in rocks where graptolites are absent. Before the end of the Devonian the ammonites had evolved, and they are probably the single most important zonal group from the Carboniferous through to their extinction at the end of the Cretaceous, although their use is often supplemented by other invertebrate organisms. Micropalaeontological zonations become progressively more important through the later part of the Palaeozoic and the Mesozoic, and the use of small, unicellular microfossils known as foraminifera outweighs that of other kinds of fossils in the Tertiary period. Zones are capable of dividing time into fine slivers. It is estimated that at best a zone (or its ultimate subdivision, a subzone) can divide time into slices of about half a million years or even less, a remarkably precise calibration.

RADIOMETRIC AGES

Obtaining an age for rocks in millions of years is now a routine geological method, and provides more accurate data which help answer the question of how long the processes shaping the Earth have taken. The method of determining such a radiometric age depends on the varieties of elements known as radioactive isotopes. These are unstable forms of elements that decay (change) slowly into other elements, in the process giving off radioactive emanations like gamma rays. The rate at which this happens is

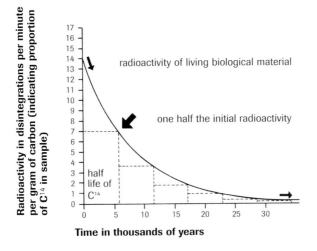

left: Time measured by radioactive decay. The half life is when half the original radioactive element has decayed away.

controlled by a simple natural law, so by knowing the amount of the parent element left, and the daughter element generated, it is possible to calculate the time taken to produce what we see today. The radioactive clock, unlike fossil clocks, converts immediately to millions of years. The method depends on accurate measurement of numbers of atoms, and techniques for doing this have improved markedly over the years, thus making the results ever more reliable. The radioactive clock is often set to zero when a liquid rock cools below a certain point. Many radiometric ages are obtained from igneous rocks, which of course, do not contain fossils, so the method often operates where fossils cannot be of direct use.

The methods originally employed depended on a few of these natural transformations, particularly entailing one isotope of uranium changing to another of lead. More and more methods are in use today, and these can now be used on sedimentary as well as igneous rocks. They are often referred to by the names of the mother and daughter elements, such as rubidium–strontium (Rb–Sr) dating, or potassium–argon (K–Ar) dating. The rates at which the decay occurs vary from one natural reaction to another; the slower the process the more use will that reaction be for dating extremely old rocks. Radiocarbon dating, involving the change from carbon-14 to nitrogen-14 happens relatively quickly (geologically speaking), and this method is useful for dating Pleistocene and younger events. This method, particularly with modern refinements, can give a resolution which exceeds that of any other technique, and is often the only recourse for dating isolated sites (like the abandoned camp fires of our forebears). At the other end of the scale the application of radiometric dating techniques has revolutionized our understanding of the huge areas of Precambrian terrain from which fossils are unavailable. The dating of meteoritic and Moon rock has provided the probable age of the Earth.

Fossils are best utilised for relative dating in the time period between the Precambrian and Pleistocene, and this is the timescale with which this book is primarily concerned. Here ages derived from radioactive decay of elements provide a scale into which the stratigraphic ages of fossils can be linked. Radiometric ages always carry a margin of error ($\pm$) of some magnitude which is diminishing as methods become more accurate. The relative timescale provided from fossil evidence is still useful, especially since most rocks cannot be dated radiometrically, and the use of fossils is not likely to be replaced by absolute ages. The two methods are complementary; fossils are used in the fine division of periods, radiometric ages provide a guide to how long ago events took place. The cooling ages in igneous rocks give the time at which particular bodies, for example of granite, were intruded into the surrounding rock. Where rocks have been reheated and folded during mountain building events, this too can 'set' the radioactive clock, providing a method of determining when major phases of metamorphism occurred. All these dates can then be placed into the geological timescale, which provides the general framework for the timing of events. The combination of radiometric ages and the relative stratigraphic scale is one of the major achievements of geology.

MAJOR CYCLES IN EARTH HISTORY: MOVING CONTINENTS

The evolution of life cannot be separated from the evolution of the planet which is its cradle and its grave. It is now certain that the surface of the Earth itself has changed its configuration several times. The Earth's crust is a thin skin, of almost negligible thickness compared to the lithosphere and mantle that lies beneath it. The crust moves, and, as it does, continental geography changes. This happens at an immensely slow rate (1–2 cm or about 0.5–1 in, per year) resulting in what is often called continental drift, but cumulatively the effects are profound. For example, at present Africa and South America are moving further apart. The blocks of continental crust that form these huge areas are thick, tough, and relatively stable. The crust that floors the oceans is relatively thin and mostly composed of basaltic volcanic rock. As the continents drift apart new oceanic crust in the form of volcanic rock wells up from the mantle along the mid-ocean ridges. The Earth's crust is composed of a number of more or less rigid plates,

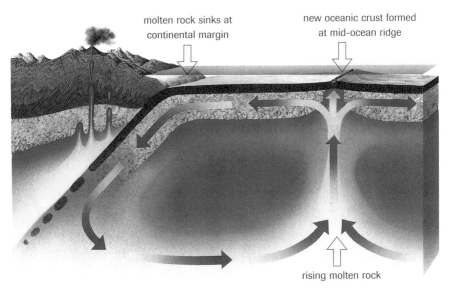

molten rock sinks at continental margin

new oceanic crust formed at mid-ocean ridge

rising molten rock

above: Generation of new oceanic crust occurs at the mid-ocean ridges, and destruction of the oceanic crust takes place at the continental margin.

drifting past, towards or away from one another. As these plates are generated along the mid-ocean ridges, so other plates have to be destroyed elsewhere. Thin oceanic crust is destroyed along subduction zones, where it plunges beneath the relatively rigid continental blocks.

An important subduction zone lies off Japan today. This area is one of sudden earthquakes, for the downward slide of oceanic crust is far from smooth. The physical expression of these downward-diving plates is found in deep ocean trenches offshore. Along the line of origin of new oceanic material, volcanoes may achieve sufficient height to break above the sea surface as volcanic islands. A different kind of volcano may erupt along the line of plate consumption, as the plates plunge down and melt at depth.

The geophysical basis of plate movements is now being fully explored, although, oddly enough, it was the supposed lack of an appropriate physical mechanism that made the scientific establishment sceptical of the whole idea of continental drift 90 years ago. As far as the palaeontological world

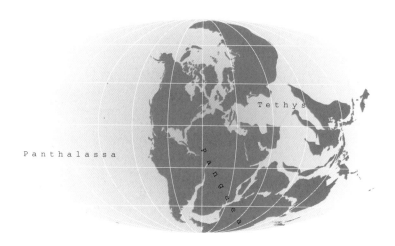

above: Reconstruction of the Permo-Triassic continent Pangaea about 250 Ma ago, which has broken up to give the present day continents.

is concerned, the evidence of fossils was one of the first compelling bodies of evidence to show that continental drift was a reality. For example, it was observed that the late Palaeozoic fossil plant floras with *Glossopteris* were remarkably similar in India, southern Africa and South America. Subsequently the same peculiar plants turned up in Antarctica. How could one explain such a distribution with the continents as they are at present? It seemed unlikely that these terrestrial organisms could have drifted across such wide ocean barriers, and the idea of thin land bridges connecting widely separated areas seemed preposterous. However, if one fits the continents back together to form Gondwanaland the *Glossopteris* floras are brought into relatively close proximity, and none of these objections apply. The flora looks like a relatively continuous belt occupying what were cooler latitudes at the time. The same arguments were applied to terrestrial vertebrate fossils. Of course animals and plants are capable of crossing bodies of water, but a look at the animals which have managed to reach remote islands today demonstrates the fact that only a few kinds of animals succeed in doing so. The oceans are effective barriers.

The combination of faunal, floral, geological, and geophysical information led inevitably to the conclusion that the present continents were fused together in one great 'supercontinent' in the Permian. This was called Pangaea. Continental drift began in the Mesozoic to break up Pangaea into the present continents. These drifted apart slowly, with widening oceans between them, a process which continues to the present day. This means that the basaltic sea floor of the Atlantic and Indian Oceans must have been created at the ocean ridges since the Jurassic. To prove this, no fossils older than Jurassic are known from the sediments which have accumulated in these regions, or indeed in any region of the sea floor. All sea floors are, geologically speaking, young rocks, and the youngest of all are still being created in volcanic eruptions like the one that built Surtsey (off Iceland) in a matter of days.

In some cases, the drifting continental blocks, or the island arcs flanking them, collide with one another. The most impressive results of this are mountain chains, like the Himalayas which were thrown up by the collision of the Indian subcontinent with the main continental landmass of Asia. Such effects are highly dramatic, but the elevation of the Himalayas (which continues) has taken millions of years, and has involved several phases. The squeezing of the rocks along linear mountain chains results in their being folded, dislocated, and maltreated in many other ways which are a delight to a geologist, but sometimes a cause of gloom for the palaeontologist, because the fossils in the rocks are subjected to the same treatment, and may emerge at the other end much the worse for wear. The whole process of squeezing culminates in heating, metamorphism and sometimes melting of the rocks, which effectively destroys the fossil record, although there are some remarkable examples of fossils having survived enormous temperature elevation.

above: When landmasses collide, large areas of the sea floor are folded and uplifted. Marine fossils, like this ammonite *Virgatosphinctes* found at an altitude of over 5000 m (16405 ft) in the Himalayas, may be raised to great heights far from their origin.

As Pangaea broke up, the dispersing continents carried the fossil remains of the Permian–Triassic supercontinent to their present scattered positions. The animals that lived on the drifting continents thus became isolated from their contemporaries in the rest of the world. For terrestrial vertebrates such isolation can result in the marooned animals having their own independent evolutionary history. Australia separated from Pangaea and carried eastwards a cargo of early marsupials which eventually became isolated from further contact with the more advanced mammals that came to dominate the rest of the world. The early migration routes into Australia were through South America and Antarctica. The eventual isolation did not stop evolution, quite the reverse, the marsupials evolved into a very varied group of animals. They were able to occupy almost all the ecological niches that were available to them, from burrowing, tree climbing or grazing to carnivorous or scavenging habitats. The primitive pouch is a common feature linking all of these animals, from the marsupial mouse to the great grey kangaroo. Some extinct giant marsupials died out after humans first settled in this self-sufficient world. The subsequent introduction of the domestic cat has done a lot of damage to smaller marsupials; it does not take long to undo what tens of millions of years of plate tectonics have created. South America was similarly isolated until the recent geological past and another set of endemic mammals evolved, including the giant sloth (*Megatherium*) the bones of which Darwin collected in his voyage on the *Beagle*.

Plate tectonics has also had its effects on marine organisms. The free floating larvae of most marine organisms mean that oceans are not the barriers for sea creatures that they are for marsupials. Marine animals are, however, adapted to particular water temperatures, which is why the species of molluscs in the tropics are generally different from those found in shelf seas that surround the North Atlantic. The distribution of marine fossils can be used as a kind of thermometer to show how the water temperatures have changed as the continents, with their fringing seas, have changed position relative to the lines of latitude. In some cases the continents themselves act as a barrier for the marine animals. The general North–South direction of the Americas and Africa effectively isolates the Atlantic Ocean from the Indo-Pacific today, which has resulted in species endemic to each region. The opening of a seaway connecting separate oceans quickly results in mixing of faunas, a process which we have been able to see in action in the short time since the Panama Canal was opened.

In the last 30 years there have been attempts to trace the history of continental distribution back still further. Why should we suppose that plate movements only started with the disruption of Pangaea? It is now certain that Pangaea itself was only a phase in the development of the face of the Earth. The supercontinent was the product of an earlier phase of drifting that brought together other, separate plates, so that in the Ordovician the world had a number of dispersed continents, as we do today. Since the process of continental drift continued far back into the Precambrian, it is a difficult business to reconstruct the patterns of very ancient continents and seas that have long since disappeared into subduction zones. However, the mysteries of these distant times are starting to be unravelled; geophysicists can now identify at least three ancient 'Pangaeas' in the Precambrian, each separated by another phase of dispersed continents. The Earth has been redesigned several times.

MAJOR CYCLES IN EARTH HISTORY: FLUCTUATING OCEANS

Fossils of marine animals can now be collected from rocks covering the far interior areas of continents which have been stable blocks since the Precambrian, regions which are now far removed from the oceans. In some cases marine deposits of this kind are explained by the fact that regions now isolated from the open ocean were formerly at the edge of continents when they were separate drifting plates. The Himalayas, the origin of which was described above, are an example where sediments that accumulated at the edge of the drifting Indian continent, and Asia, were sandwiched between the two continental blocks as they collided. There is no problem in explaining the presence of thick marine deposits here, the former sites of

above: Flat lying sediments like these in the Grand Canyon in the interior of North America indicate that in the past the sea extended over the continent.

seaboards that have been crushed during continental collision. In other cases, such as the interior of North America or Australia, there have not been comparable continental collisions and yet the sea evidently flooded over the continental interior from time to time, leaving deposits containing marine fossils extending deep into the heart of the continents. The only explanation for these kinds of deposits is that the sea has periodically extended much further over the continental areas than it does today.

These periods of flooding (or marine transgressions) are now recognized as one of the most important cycles of physical events that have affected the Earth, and their effects on the course of evolution have only recently been considered. Major marine transgressions of this kind have occurred at intervals throughout geological time. Since they are due to the influence of rising sea level, they are simultaneous over the whole world. They afford another way of subdividing geological time. When the sea extended over the widest areas, naturally the deposits of that age are the most widespread and tend to be the most well known. Conversely, there were periods when the sea drained off the continents (regressions); at these times marine deposits were confined to areas peripheral to the continents and in the open oceans, while terrestrial sediments extended over the areas where marine deposits had accumulated during the transgressive phase.

These oscillations may be related to continental movement. Some scientists have suggested that the periods of transgression correspond with active phases of generation of new ocean floor at the mid-ocean ridges, and the regressive phases correspond with periods of temporary standstill of drifting activity. Another cause that has been invoked are the Ice Ages. During a time of ice advance an enormous quantity of water is locked up in ice sheets, and world sea level falls as a result. As the ice melts, sea level rises, producing a transgression. Rising and falling sea levels in the last 1 million years of the Pleistocene Ice Age were certainly controlled by glacial events. Whatever the cause, these great cycles have influenced the spread and evolution of life, both on land and in the sea. During the transgressive phase shallow marine faunas are widespread and diverse; in tropical waters reefs build up, and support some of the most diverse communities of marine animals. Sudden changes in conditions were one of the factors leading to marine extinctions.

MAJOR CYCLES IN EARTH HISTORY: FLUCTUATING CLIMATES

As we have seen, the continents are constantly on the move, the seas may advance upon them and drain off again. This may already seem a world in which everything is in a state of flux, but to provide the environment in which past organisms lived, we have to introduce one more cycle of change. The world climate has varied, from generally warm to glacial phases. This has to be distinguished from changes of climate on any one continent, i.e. if a continent is moving under the influence of plate tectonics it may pass across climatic belts, starting in the tropics and finishing up near the pole (like the southward drifting Antarctica). Climatic cycles involve climate change over the Earth as a whole. Such changes are related to global sea level changes, as the major glacial episodes will result in a regressive (draining) cycle over the continents.

There have been at least four major periods of glacial activity in the period over which life has left its abundant traces in the rocks. The first of these was in the late Precambrian (750 Ma ago) (and there were almost certainly others in the further reaches of Precambrian time), somewhat before the massive diversification of animals with hard parts at the base of the Cambrian. According to some scientists this was the biggest glaciation ever, known as 'snowball Earth'. It is certainly true that glacial influence extended all the way to the tropics at this time, but whether the Earth was entirely frozen is debatable. The same scientists link the end of the frigid conditions with the subsequent explosion of life. The second glacial phase occurred late in the Ordovician when much marine life was exterminated. The third and probably the most important glacial was the Carboniferous–Permian event, or rather events, the effects of which are preserved over a huge area of the Southern

above: This specimen, known as tillite, is the consolidated reamains of the rock fragments and powder that are left behind as a glacier melts and retreats. This tillite from Kimberley (West Griqualand, South Africa) provides evidence of much cooler conditions on a continent which today has a climate too warm for glaciers to form

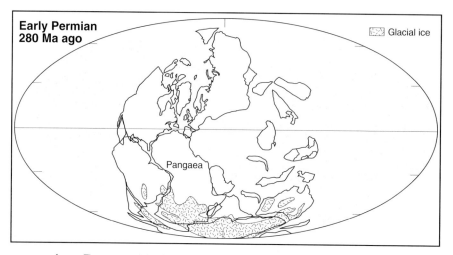

above: The extent of the Carboniferous-Permian glaciation in Pangaea.

Hemisphere. The distribution of the glacial rocks here was used by Wegener in 1915 as one of the arguments in favour of continental drift. The effects of this glaciation in the Southern Hemisphere were also felt in Carboniferous Europe, which at that time lay near the Equator. Glacial retreats released immense quantities of water, which resulted in sea level rise. On some occasions the rise was sufficient to drown the coal measure swamps, inundating what had formerly been luxuriant jungle, and interpolating marine fossils in terrestrial rock successions. This late Palaeozoic glaciation lasted for a long time; between the major pulses of glaciation in the Carboniferous and the mid-Permian there was an interval of approximately 20 million years.

We are currently within a glacial phase; the Pleistocene epoch, extending back over the last 1 million years. Over this period, radiometric dating techniques combined with careful analyses of fossil pollens and various microfossils, have enabled a subdivision of geological time on a much finer scale than for the earlier parts of the stratigraphic column. The story so revealed is very complex. The ice has advanced and retreated numerous times; the original idea of four great advances has been shown to be a gross

above: Glacial 'pavement' with the scratch marks made by an overriding ice sheet. Such marks can be found where there are sites of former glacial activity.

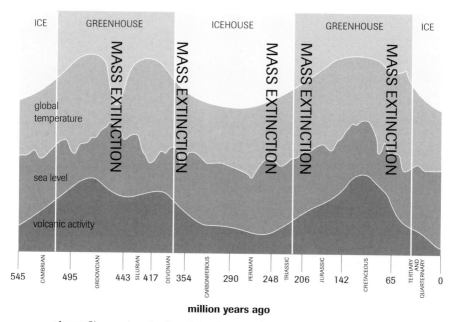

ICE | GREENHOUSE | ICEHOUSE | GREENHOUSE | ICE

MASS EXTINCTION MASS EXTINCTION MASS EXTINCTION MASS EXTINCTION MASS EXTINCTION

global temperature

sea level

volcanic activity

545 | CAMBRIAN | 495 | ORDOVICIAN | 443 | SILURIAN | 417 | DEVONIAN | 354 | CARBONIFEROUS | 290 | PERMIAN | 248 | TRIASSIC | 206 | JURASSIC | 142 | CRETACEOUS | 65 | TERTIARY AND QUARTERNARY | 0

million years ago

above: Changes in volcanic activity, climate and sea level influence each other and may cause mass extinctions.

underestimate. The number of glacial phases is now thought to be approximately 15. Phases of retreat between ice advances included times when the climate was very much milder than today. There is no reason to suppose that the process is at an end; another ice advance may be on the way in the future. The huge quantities of water locked in the present ice sheets mean that the area of exposed continent is probably greater now than it has been for most of the Tertiary.

Between glacial phases there were other periods in which the climate of the world, as recorded in the rocks, was generally warmer and more equable. The Silurian and much of the Cretaceous were times of unusual warmth. At these times the tropical zones expanded to cover a greater area of the oceans and continents. Wide spreads of warm-water limestones are typical of the rocks deposited at such times, and with the limestones are found fossil animals adapted to the warm oceans.

plate 1 (above) Hexactinellid sponge *Coeloptychium agaricoides*, Cretaceous, Westphalia, Germany. These two specimens were extracted from a matrix of white chalk, hence their colour. The cap of the specimen is 8 cm (3.2 in) across.

plate 2 (left) Fossil glass sponge, *Hydnoceras tuberosum*, Devonian, New York, USA. *Hydnoceras* is a Devonian and Carboniferous genus, but glass sponges of similar form, without knobs, go back to the Cambrian. Diameter is 20 cm (8.0 in).

plate 3 Rugose coral section, *Cyathophyllum* sp. preserved in limestone, Devonian, Devon, UK. The gaps between the septa of the coral skeleton have filled with calcite, showing as the lighter colours. Diameter 6 cm (2.4 in).

plate 4 *Thamnopora cervicornis*, Devonian, Devon, UK, approximately 370 Ma old coral.

plate 5 Tuning-fork graptolite *Didymograptus murchisoni*, Ordovician, Wales, UK, flattened and preserved on the flat bedding surfaces of a black shale. Individual specimens grow to a length of 5 cm (2.0 in) or more.

plate 6 *Dichograptus octobrachiatus*, Ordovician, found in eastern United States, Texas, Britain, Canada and Australia.

plate 7 (above)
Three brachiopods from the Carboniferous of Castleton, UK. *Spirifer striatus* (left), *Antiquatonia hindi* (top right), and *Echinoconchus punctatus* (bottom right).

plate 8 (right)
Slab containing many Silurian brachiopods of *Camarotoechia* sp.

plate 9 Tubes (3 cm or 1.2 in long) of polychaete worm *Rotularia bognoriensis*, from the Eocene of Bognor, UK preserved in siltstone. The irregularity of the tubes, which become straight near their apertures, distinguishes them from snail shells.

plate 10 (above left) Trigoniid bivalve, *Scabrotrigonia thoracica*, Cretaceous, Tennessee, USA. The radial ribs 'chopped up' into little knobs are characteristic of the trigoniid bivalves. The longest diameter of this specimen is 5.5 cm (2.2 in).

plate 11 (above) Fossil sundial shells, *Architectonica millegranosa*, Pliocene, Oriciano, Italy. This species is one of a fairly large genus that survives today. Diameter about 3 cm (1.2 in).

plate 12 (left) Fossil oysters, *Gryphaea arcuata*, about 200 Ma old, found in the Lias clays and limestones of the Jurassic period. Commonly called Devil's toenails.

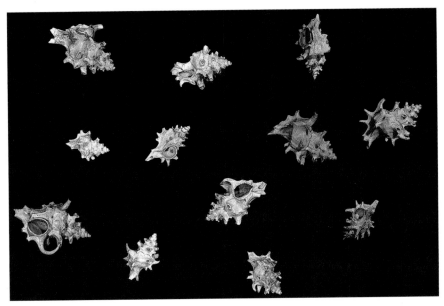

plate 13 *Typhis pungens*, fossil gastropod, Eocene. The fossils here illustrate the different sizes gastropods of one species attain. *Typhis* still lives today, e.g. in the seas around Japan. The largest is 3 cm (1.2 in) long.

plate 14 Eocene gastropod, *Voluta muricina*, Epernay, France. The species is distinguished by its tall spire, elongate aperture, prominent spines, but without the spiral ridges seen on many species. Length 7 cm or 2.8 in.

plate 15 Gastropod, *Harpagodes wrightii*, Jurassic, Gloucestershire, UK. This large gastropod (nearly 15 cm or 6.0 in) has long, stout spines which are an unusual feature, discriminating this species from other gastropods.

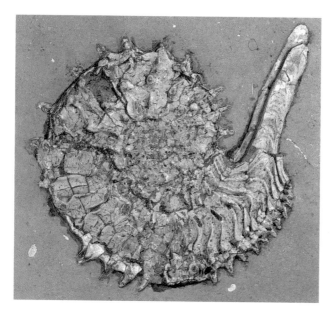

plate 16 Severely flattened ammonite, *Kosmoceras acutistraitum*, Jurassic, Wiltshire, UK. This species is particularly distinctive because of the extended flanges at its apertural margin (lappets). Longest diameter here is 9 cm or 3.6 in.

plate 17 Jurassic ammonite, *Promicroceras planicosta*, Somerset, UK. Characteristics of this species include the small size, and unbranched and strong ribs, giving the shell its ramshorn appearance. Individuals are 2 cm (0.8 in) in diameter.

plate 18 A section cut and polished through Jurassic nautiloid, *Cenoceras pseudolineatus*, from Dorset, UK. Shows the internal chambers filled or partly filled with calcite. Diameter 7 cm (2.8 in).

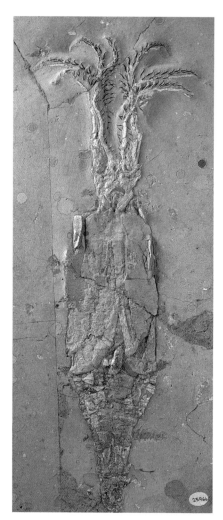

plate 19 (left)
Belemnotheutis antiquus,
Upper Jurassic, Malford,
Wiltshire, UK.

plate 20 (right) A group of
belemnites, *Acrocoelites
subtenuis*, Jurassic, Yorkshire,
UK. There are very many
belemnite species in Jurassic
and Cretaceous rocks.
The longest specimen here
is 9 cm (3.6 in) long.

plate 21 Fossil crinoids, or sea lilies, Lower Jurassic, Germany. Their modern relatives include other echinoderms, such as sea urchins, starfishes and sea cucumbers.

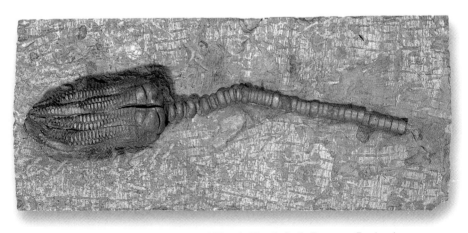

plate 22 Triassic crinoid, *Encrinus liliiformis*, Muschelkalk, Germany. *Encrinus* is typical of marine limestones of the Triassic of Europe. Length 18 cm (7.2 in).

plate 23 (right) Silurian cystoid, *Pseudocrinites magnificus*, UK. The fossil is composed of individual calcite plates. This rare and peculiar fossil looks rather like a sea-lily without arms, and is 5 cm (2.0 in) long.

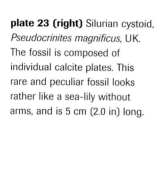

plate 24 (above) Fossil starfish, *Archastropecten cotteswoldiae*, Jurassic, Gloucestershire, UK. The specimen occurs in a fine-grained limestone, which fortunately splits easily around the enclosed fossils. This species has a small central disc, and long, slender arms.

plate 25 (right) Starfish, *Platanaster ordovicus*, from the Ordovician of Shropshire, UK.

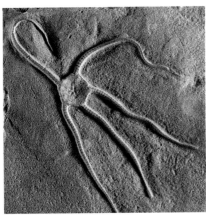

plate 26 (above) Jurassic sea urchin, *Hemicidaris intermedia* , Wiltshire, UK.

plate 27 (left) *Palaeocoma egertoni*, Jurassic, Dorset, UK, 195-189 Ma old brittle-star.

plate 28 (above) Cidaroid sea urchin, *Plegiocidaris coronata*, Jurassic, Ulm, Germany, preserved in a fine-grained limestone in full relief. The spines are not preserved on this specimen. Diameter 5 cm (2.0 in).

plate 29 (right) Sea urchin, *Tylocidaris clavigera*, Cretaceous, southern England, UK. Preserved in its original calcite in a matrix of chalk, this specimen is unusual because it has the spines still joined to the rest of the animal.

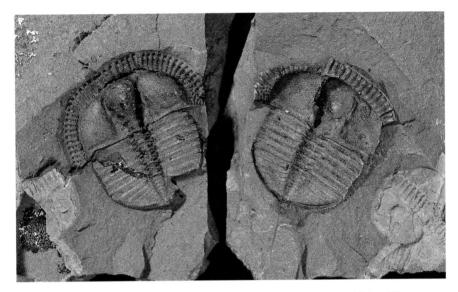

plate 30 Trinucleid trilobite, *Trinucleus fimbriatus*, Ordovician, Builth, Wales, UK. These trilobites are blind, with an inflated mid-part of the head. Around the head perimeter there is a pitted fringe, unique to this kind of arthropod.

plate 31 Trilobite, *Dalmanites myops*, Silurian fossil trilobite from the Wenlock Limestone, Worcestershire, UK. Specimen measures 4 cm (1.6 in) head to tail.

plate 32 (left) Devonian trilobite, *Phacops rana*, Ontario, Canada. The mid-part of the head is covered with coarse tubercles and expands forwards, and the eyes include a few, very large lenses.

plate 33 (right) *Eurypterus lacustris*, Silurian, Buffalo, New York, USA. About 440-417 Ma old. Extinct predatory arthropod, length 14 cm (5.6 in).

The climatic changes themselves are partly under the control of atmospheric conditions. The warmest phases tend to be 'greenhouse worlds' when greenhouse gases, especially carbon dioxide, reach high levels. 'Icehouse worlds' occur in times when carbon dioxide levels are low. It is becoming apparent that we are moving into a greenhouse world induced by our own excessive burning of fossil fuels. The outcome may well control the future of our species.

THE CHANGING WORLD

The arrangement of the continents changes, the sea levels change, and the world climate changes. This is the shifting stage on which organic evolution has acted and will continue to do so. The cast of characters has changed repeatedly; the animals or plants adapting to the conditions pertaining while they lived. In this dynamic world, what are we to make of James Hutton's ambition to interpret the geological past by processes that operate at the present day? The basic assumption still stands, that though the stage may have changed, the processes that shaped the scenery were the same in the past as they are at the present. Physical laws do not change. It is only by a thorough understanding of the forces at work today that the past can be reconstructed. Life is bound up with the story of the changing Earth, and it is foolish to pretend that the history of life can be fully understood without its dynamic setting. The history of life has been so closely bound up with the history of our planet that it is likely that some small change in that history would have produced a change in the course of evolution. If the climate had not changed in Africa a few tens of thousands of years ago, would modern humans have evolved? There is nothing inexorable about the course of evolution; rather it is a complex *pas de deux* between the changing environment, and the capacity of organisms to respond to those changes.

Chapter Three

ROCKS AND FOSSILS
· ·

A VISITOR to a museum will see perfectly preserved and spectacular fossils neatly arranged in glass cases. It is easy to forget that these fossils were found by cracking open rocks. All fossils are found in rocks that were originally unconsolidated sediments, and the study of fossils can be enhanced by knowing about the rocks that enclose the fossils. Certain environments which today support a rich and varied plant and animal life have no sediments forming in them, and the organisms living there have virtually no chance of being preserved in the fossil record. Mountainous regions, for example, are dominated by the erosion of the rock forming the ranges, and therefore no permanent sediment is formed there. Torrential rain and rapid weathering, aided in some climates by the action of frost, rapidly destroy much of the organic material, and the chances of any preservable remains reaching a lowland river where permanent sediment is accumulating are remote. As the fossilization potential of a mountainous environment is low, the faunas and floras of mountainous regions of the past are most unlikely to be represented in the fossil record.

The study of fossils is connected with a suite of rocks that are formed in environments where sediments accumulate, and have a high chance of becoming rocks. Such environments cover a large part of the surface of the globe, including most of the submarine areas, and some of the lowlands, where rivers and lakes accumulate sediments of many types. This chapter is concerned with the different sites in which sediments may form.

In some instances the site of sediment accumulation is a direct reflection of the environment in which an animal or plant lived. Lake fishes and plants, for example, are to be expected in lacustrine (deposited in a lake) sediments. The wide variety of sediment sites will be reflected in an equal variety of animals from different habitats. Extensive studies of recent sediments enable the interpretation of sediments from the past. Most sedimentary rocks retain in fine detail the features acquired while they were accumulating. By studying recent sediments it is possible to determine the site of deposition of past rocks, and from this to understand more about the environment in which the fossil animals in the rock lived.

Occasionally dead animals and plants travel for long distances before finally becoming entombed in the sediment. Empty shells of *Nautilus* have been found over a much wider area than that in which the animal lives. Drifting logs can be found hundreds of kilometres from land, and when these become waterlogged, they sink and eventually become incorporated into the sediment.

SEDIMENTARY FACIES

The rocks formed at a particular site, each with their own peculiar characteristics, are called sedimentary facies. Just as at present, sediments of many different facies are accumulating in different places, so in the past rocks with totally different appearances may have accumulated at the same time in different environments. The fossils found in such rocks also differ from one site to another, as they are also related to the environment in which the sediment accumulates; different sedimentary facies may have different assemblages of fossils. The term facies fauna is applied to an assemblage of different fossils that are found together in one particular sediment type. Not all animals are so limited: in the sea, for example, many free swimming or floating organisms are independent of the conditions of sediment accumulation on the sea floor.

The diagram below shows the main sedimentary facies in a hypothetical section running from the mountainous interior of a continent in tropical latitudes to the open ocean. In the lee of the mountain range a rain-starved desert accumulates mostly wind-blown sand derived from the weathering of steep buttes. Occasional torrential bursts of rain produce flash floods, which sweep down the steep walled valleys (or wadis) carrying with them all the weathered material, which spreads out into broad fans. The flood water drains into temporary pools, which evaporate rapidly in the hot sun, but sometimes linger long enough to allow brief bursts of specialized

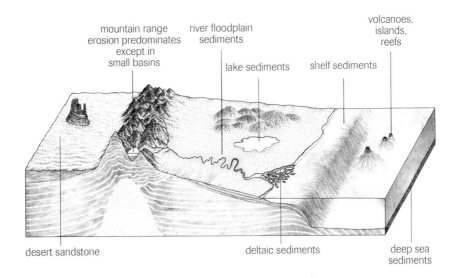

above: A section from a desert area through a mountain range, across a sedimentary plain to the shallow sea, and ultimately to the deep sea, showing the different sites in which fossils may accumulate.

animals to colonize the warm water. As the waters evaporate, mineral salts are concentrated within the pools, often crystallizing out as the drying continues to produce white, glistening salt pans. Most of the animal and plant life here has to be very specialized to cope with the harsh conditions.

On the mountains themselves erosion predominates, aided by the action of ice found on the higher peaks. Frost shattering helps to splinter rocks into shards and blocks that tumble down slopes and become the raw material for streams to move inevitably downwards towards the sea. Melting snow produces raging torrents with immense transporting power, that in full spate can move and break huge blocks of rock into smaller cobbles which are more easily transported. Little sediment accumulates here except in deep depressions between ranges (inter-montane basins), the scree slopes at the edge of the mountains, and in the deposits of streams and rivers. Much of the sediment produced on the mountain sides becomes ground into smaller particles. In this state it can be transported by the large and more sluggish rivers through the foothills and beyond. Occasional floods originating in the mountains may transport huge amounts of material, often with disastrous consequences such as flooding on the plain.

The deposits associated with major rivers are silts, clays and sands, often with characteristic combinations of sedimentary structures that reveal their fluviatile origins. Large lakes lying on the plains are important and potentially fossil-bearing sites of sediment accumulation. In low lying areas swamps support prolific vegetation which decays to form beds of peat. Insects and other animals adapted to this habitat may be destined to be preserved therein.

As the rivers wind towards the sea, their floodplains broaden and they meander over the plains, breaking their banks and readjusting their courses at times of flooding. They can carry huge quantities of sediment, mostly in suspension (at the present day about 10,000 million tonnes of sediment finds its way into the seas over the whole world each year) but also in solution. At the junction of land and sea the river begins to shed its load, partly from the effect of fresh water meeting salt water, and partly because its energy and carrying power dwindles. Typically the river builds a delta, and the silts, sands and clays of deltas are one of the most important sedimentary facies in the fossil record. Swamps may form around the small streams (distributaries) that criss-cross the delta, and there are many habitats suitable for the successful life of animals and plants, as well as their rapid preservation on death. Deltas slowly build out into the sea, forming an advancing wedge of sediment that may extend for many kilometres. In such cases the seaward edge of the delta is younger than its fossilized edges preserved on the landward side. Deltaic sediments, although in geographic continuity, are not all the same age. These are diachronous deposits.

Smaller rivers, like the Thames, often do not build deltas but enter the sea in estuaries, where salt and fresh water oscillate in influence in tidal stretches of water. Estuaries also produce characteristic sediments, and have specific sets of organisms associated with them.

We now pass into the marine environment, which has produced the bulk of the sedimentary rocks and in which the record of past life is most complete. In many areas the contact between land and sea is erosional, as any visit during a storm to a resort with sea cliffs will prove. Storm beaches of rounded cobbles often form the boundary between the sea and the land, and may be preserved in the rocks as coarse conglomerates. Further offshore, sediments are generally finer sands, silts and clays, and far more suitable for preserving fossil remains than the deposits of storm beaches. The finer material from rivers and from the direct erosion of the land by the sea contributes to the sediment deposited on the sea bed. Powerful currents

above: Fossil ripple marks preserved in sandstone. The detailed structure of the ripple marks can reveal much about the past sedimentary environment.

on the sea floor often influence the pattern of distribution of such sediments. Shallow marine deposits also have characteristic sedimentary features; the ripple marks of tidal seas are frequently encountered in rocks, and the tracks of animals can be preserved here. Fossil footprints in sand are no doubt preserved somewhere today, recording the holiday habits of humankind. Although there are many exceptions, the grain size of the sediment generally decreases away from the coast, so that deposits further offshore tend to be fine muds that will produce clays or mudstones in the geological column.

In warm climates, where there is not much land-derived sediment available, other forms of sedimentation become important, especially the formation of limestones. Many limestones are formed from the consolidation of lime muds, the precipitation of which is mediated by bacteria. The lime in these muds is a form of calcium carbonate (aragonite) with minute crystals, which can only be deposited in warm and shallow seas. Where there is turbulent agitation of the water in shallow areas, the carbonate may be laid down in concentric layers around tiny organic nuclei to produce rounded ooliths. A rock largely composed of ooliths is an oolite. In spite of their rather specialized mode of formation oolites can form a surprising volume of limestone formations, covering hundreds of square kilometres in the Ordovician rocks of North America, and almost as extensive in the Carboniferous and Jurassic rocks of Europe. Numerous species of animals are confined to a lime substrate, and naturally one finds their extinct counterparts preserved in limestones. Some limestones are composed very largely of the remains of calcareous animals.

above: An oolitic limestone (Rutland, UK) showing the perfectly rounded ooliths of which it is composed. Such limestones formed in shallow, agitated marine conditions, and only in warm climates.

As the deep sea is approached there is generally a reduction in the amount of sediment that can reach the open ocean from land sources. On the floor of the open ocean at great depths, large areas are covered with a fine ooze, which is formed predominantly from the skeletons of minute planktonic organisms that have rained down from the surface waters. The greatest density of planktonic life exists in shallow depths where light penetrates and allows microscopic plants to live, as well as small organisms which feed on these plants; it is the shells of these small animals that form much of the deep sea sediment. The rate of accumulation of these deep sea deposits is very slow, only a few millimetres (less than 0.4 inches) per thousand years. Several kinds of single-celled organisms may dominate these deep sea oozes; the foraminiferan *Globigerina*, which has a calcareous test, and the delicate, siliceous radiolarians are especially important. These deep sea deposits record a remarkably complete history of the evolution of the planktonic organisms contained within them. Even these small shells, however, are to some extent soluble in sea water, and at very great depths pressure increases this solubility so that they do not survive. Here the only deposit is red clay (often in fact brown in colour), a sediment which accumulates extremely slowly. It is composed of the finest wind-blown dust, volcanic ash carried by winds from distant eruptions, and, occasionally, the insoluble traces of animals such as the teeth of sharks or the ear bones of whales. In these abyssal depths curious nodules with a high proportion of manganese grow slowly in the red clay, and there has recently been speculation about the possibility of exploiting these as a mineral resource, surely the least

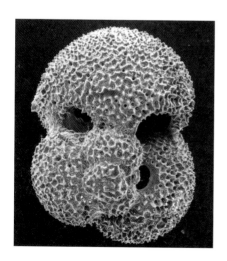 

left: The foraminiferan *Globigerinoides* which is dominant in deep sea oozes.
right: The delicate, siliceous skeletons of radiolarians such as *Acanthometra* with skeletal spicules, cover large areas of the deep sea floor.

accessible ore in the world. In spite of the inhospitable, light-less conditions in the abyssal seas there is a variety of life, and there are specialized and often bizarre fish and crustaceans that live only there. These leave little fossil record.

In the open ocean, volcanic islands form sporadic sediment sources, both from the erosion and redistribution of the volcanic rocks themselves, and because they reproduce the same sort of conditions that pertain on the continental shelves. The clear water surrounding such islands in the tropics is often suitable for the growth of coral reefs, discussed later in this chapter.

In Arctic regions the influence of ice as both an erosional and depositional agent is paramount. The scouring action of ice, using rocks enclosed within glaciers as tools to scrape and gouge the underlying rock surfaces, produces great quantities of angular detritus. Some of the rock is ground as fine as flour. At the melting edges of glaciers, or where icebergs break off from ice caps and drift into the sea, much of the material is released and becomes sediment. Such glacial deposits (till) are often a heterogeneous selection of different rock types, dumped together, with large boulders and tiny pebbles immersed alike in the groundmass of rock flour. Not surprisingly fossils are rare in these kinds of rocks. Around the edges of ice sheets mossy bogs are common and may form deposits of peat and lignite containing the remains of organisms adapted to life in high latitudes. During the last Ice Age animals often used caves as shelters or lairs, and the deposits of cave floors have proved particularly rich sources of their bones. Further away from the ice front, major rivers took away the meltwaters to the sea, and their fluviatile deposits often preserve the remains of large mammals that lived in the surrounding areas, some no doubt fatally entrapped in bogs.

Some land-derived sediments reach the deep sea by means of turbidity currents. These are slurries of sedimentary material that are flushed from shallower areas at the edge of the continental shelf, a movement often sparked off by earthquakes. Sometimes their effects can be quite catastrophic, snapping underwater cables, and they can travel extraordinary distances, up to 300 km (190 miles) or more. Turbidity currents produce a characteristic rock type in the geological record known as a turbidite. Communities of animals which live on the sea floor and suddenly have material deposited on top of them by a turbidity current can be buried in a single catastrophe.

Just as climate influences the kinds of sediment laid down, so also it is one of the most important influences on animal and plant life. Most marine organisms are zoned according to latitude, and it is possible to represent these distributions as a series of belts approximately parallel to lines of latitude (distorted by the influence of warm and cold currents). The same sort of influences undoubtedly operated in the past. During the Pleistocene, when climates oscillated over many thousands of year between warm and cold,

above: Climatic fluctuations in the Pleistocene are often recorded in cave deposits. These bone stacks built by William Beard at Banwell Bone Cavern, UK, are nearly all those of the bison *Bison priscus* and are about 70,000 years old, indicating very different climatic conditions to today.

marine and land organisms migrated backwards and forwards with the climatic shifts to keep living in the conditions to which they were adapted. Since these oscillations ran approximately parallel on land and in the sea, this provides one of the methods of subdividing the Ice Ages. Tropical faunas and floras are richest in numbers of species, and, at the other extreme, only a few hardy species of high Arctic or Antarctic animals and plants are able to cope with extreme polar conditions. Species that do adapt to polar conditions may be found in great profusion.

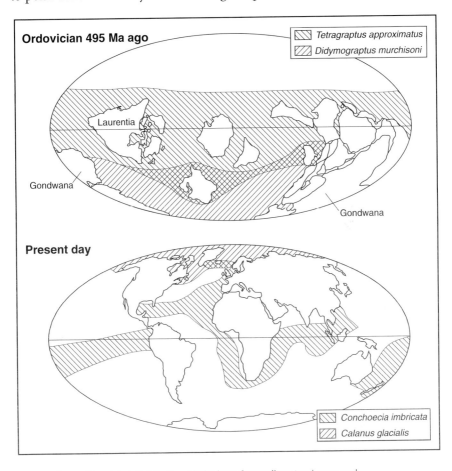

above: The global distribution of plankton (graptolites, top image and crustaceans, lower image), showing how these organisms have a latitude-controlled distribution.

We have seen how sediment type alters with the depth of water. Animals are also influenced by water depth, and it is not surprising to find different communities of fossils in rocks that were deposited at different depths. Both the sediments and the animals found as fossils within them are reflections of the past environment, and both provide clues to the life and conditions of former times.

REEFS

Organic reefs today are formed by corals and calcareous algae that make wave resistant structures, often of immense dimensions. Because the framework of a reef is composed of tough calcareous organisms which have to stand up to the buffeting of waves and the ravages of storms, they are likely to be preserved in the fossil record. Reefs can be preserved with all the constituent organisms in life position. Living reefs can only be found in warm water regions (23–25°C or 73–77°F is best) with abundant sunlight. Light is needed by the minute photosynthetic algae that live in the tissues of some reef-building corals. Reef growth can only proceed satisfactorily in salt water, and not in water of too great a depth where the vital light begins to be filtered out. Reefs cannot tolerate too much clogging sediment in the surrounding waters. The majority of fossil reefs seem to have lived under similar constraints.

The reef environment is an extremely rich one, supporting a host of animal species besides the frame builders themselves: fish, molluscs, sea urchins and crustaceans live among the corals as scavengers and predators, together with numerous encrusting organisms that benefit from the firm foothold provided by dead coral. The corals themselves feed by capturing minute planktonic food from the constant wash over their polyps. Erosion of the reef is rapid, and this builds slopes of reef waste and finer debris that can also be found fossilized. This produces a rock that is almost entirely made of fossils. Charles Darwin was the first to demonstrate that the curious

top (left and right): Fossil coral *Actinocyathus floriformis* (order Rugosa) from Carboniferous coral rich beds (Shropshire, UK).
bottom (left and right): Recent coral *Favites pentagona* (order Scleractinia) from a coral reef along the coast of Somalia.

circular pattern of atoll reefs was due to the foundering of volcanic islands, around which reefs had initially formed as fringing structures. The rapid growth of the reef can keep pace with the sinking of the island, producing in the process a great amount of sedimentary waste that preserves the earlier, fossil history of the reef.

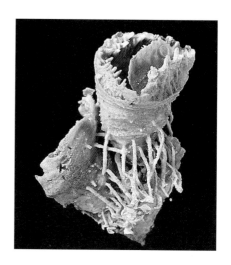

right: A conical richthofenioid brachiopod with 'lid', *Prorichthofenia permiana* from Hess Canyon (Texas, USA).

The reef as a structure has a long history, extending back even to a time before the corals themselves had evolved. Other organisms were able to adopt the same role as the corals, often assuming similar general shapes, even though unrelated in a zoological sense. The earliest reef structures are present in early Cambrian rocks in Labrador, where the frame was built by enigmatic sponge-like organisms called archaeocyathids. Corals themselves, although belonging to forms unrelated to living species, started to achieve prominence in the later Ordovician, and by the Devonian large reef structures were present. The storm-facing surface of the reefs was often formed by massive stromatoporoids, a group of sponges with only a few inconspicuous relatives living today. Fossil reefs are found almost worldwide, reflecting the warmer temperatures globally at this time. Some of the corals from these reefs closely resemble their living counterparts although this resemblance is due to similar life habits, and their detailed structure is quite different.

In the Permian rocks of Texas and adjoining states, large reef structures have been found where sponges and bryozoans were the important frame builders, along with the ubiquitous algae. The same habitat supported one

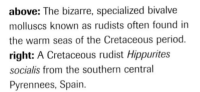

above: The bizarre, specialized bivalve molluscs known as rudists often found in the warm seas of the Cretaceous period.
right: A Cretaceous rudist *Hippurites socialis* from the southern central Pyrennees, Spain.

of the most bizarre brachiopods, a conical form with a 'lid' quite unlike the usual brachiopods. At the end of the Permian most of the reef-building organisms that built large reefs in the Palaeozoic became extinct. By Jurassic times coral reefs were again being constructed, this time by the distant relatives of the corals and other organisms that build reefs at present. The corals of Palaeozoic age are distantly related, if at all, to those of the Mesozoic to Recent, but they both built reefs of similar construction.

During the Cretaceous an extraordinary group of molluscs acquired the habit of building sea-floor structures, although they were not really reefs. These were the rudists, which again are conical, with a lid, at first glance much like the aberrant brachiopods of the Permian. However, the rudists were quite unrelated to the brachiopods, and were derived from clam-like molluscan ancestors. The rudists did not survive the Cretaceous 'greenhouse world', and were confined to very warm limestone seas; they may have been adapted to higher temperatures than pertained over much of geological history. They are important fossils in the Alps and in North America, where they form masses lying on the ancient sea floor .

During the Tertiary, the break up of the continents, with the associated volcanic activity creating islands in the oceans, permitted the establishment of the ancestral reefs that continue to be built until the present day.

The reef environment is important because it shows how different organisms can assume a similar superficial appearance when they adopt similar life habits. There is more than a passing resemblance between some of the archaeocyathids of the Cambrian and the Permian brachiopods, and some of the corals have comparable forms as well. On present day reefs more densely branched forms tend to be found on the exposed, seaward flanks, while the backreef areas have organisms with loosely branching antler-like growths. There may be variation in branching habit even within a single species, according to the site in which it flourishes. Through geological time various organisms have played the same ecological role, and the result has been similar shapes. This is an ecological control that can act on very different starting material (like bivalves and brachiopods) and produce a superficially similar end-product. The biological prerequisite of most of the organisms mentioned here is that they should be suspension feeders. Conversely, for a highly adapted organism like a reef dweller it is necessary to look carefully at the fine structure to determine the biological affinities of the organism, and not be misled by superficial resemblance.

DEEP SEA DEPOSITS

The peculiar sediments formed in the deep sea, beyond the edge of the continental shelf, also have a long history. Deposits laid down in the open ocean occupy a large area of the globe today, and there is every reason to suppose that this was true far back in the Palaeozoic. However, the area occupied by oceanic sediments in regions of Palaeozoic rocks is not commensurate with this former extent. This is because most of the ancient oceanic sediments have been destroyed where they plunge down the

above: Corals forming on a reef at Kimbe Bay (Walindi, Papua New Guinea).

subduction zones at the consuming edges of plates; oceanic sediments are the ones that disappear forever. The crust of the Earth is in a state of dynamic equilibrium, with oceanic crust being created at mid-ocean ridges, only to be consumed at the edges of plates. The break up of the supercontinent Pangaea during the Mesozoic resulted in both the creation of oceanic crust that floors the oceans today, and the destruction of previous oceanic crust, so that at present the oldest oceanic crust is probably only Jurassic in age.

In spite of the odds against it, some ancient oceanic deposits are preserved, but only in special circumstances. Sequences of turbidites form prisms of sediment, sometimes thousands of metres thick, at the edges of the continents. During phases of collision these sediments become squeezed between the colliding plates as the oceanic crust itself is consumed. The sediments respond to this pressure by crumpling, shattering, and gliding into great sheets that move away from the centre of pressure. Many of these sediments become heated, partially melted, or so contorted by pressure that any fossils they once contained are transformed beyond recognition. Some, however, survive with their fossils intact, although the strata from which they have to be recovered are almost always vertical, rather than in their original horizontal attitude, and frequently heavily distorted, with the fossils they contain distorting along with the rock. The fossils are not usually found in the turbidites themselves, but in interbedded shales, representing the quiescent conditions between turbidite slumps. In some cases slices of oceanic rocks, instead of being consumed, have been thrust beyond the danger zone, often carrying on their backs a skin of sediment. These slices have a characteristic combination of volcanic rocks (often with serpentine) with cherts and sometimes black shales. These remnants of former oceans are known as ophiolites. Where ophiolites are found in mountain belts this indicates that oceanic rocks have been obducted in that region. Adjacent areas of folded or metamorphosed rocks were produced by continent colliding with continent or island arcs docking, much of the intervening ocean which was originally present having dived to oblivion.

In order to find deposits laid down in the ancient deep seas and the fossils they contain, the right geological setting must be found. It is of no use looking in the sediments formed on the ancient shield areas of Precambrian rocks, which even in the Palaeozoic were covered by relatively shallow seas. In the areas of intensely folded rocks which probably accumulated at the edge of former continents, the chances of finding deep sea sediments are increased.

In the Cambrian, deep sea shales between turbidite sequences have been found to contain trilobites, minute, blind forms, known as agnostids. The likelihood is that these specialized trilobites were free swimming or planktonic forms, and they did not actually live on the sea floor. In the Ordovician and Silurian drifting colonial organisms known as graptolites usually dominate any assemblage of fossils recovered from between the turbidites, their remains often forming matted clots. Clumps of graptolites,

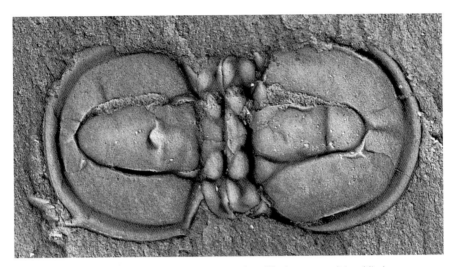

above: Agnostid trilobite of Cambrian age, found in deep water. It is a blind trilobite of only a few millimetres long, with only two thoracic segments.

perhaps killed due to a plankton 'bloom' consuming the available oxygen, slowly fell through the water column, eventually to lie down on the soft muddy sea bed; the arrival of another turbidite flush would make their entombment complete. The occasional glass sponge may have actually lived on the deep ocean floor; they are found sporadically through the fossil record in these sorts of sediments, and are still relatively prolific at the bottom of the oceans today. Special extraction techniques can recover the remains of single celled plants from marine deposits of Palaeozoic (or even Precambrian) age. These show that photosynthetic algal plankton, the basic link in the food chain of past seas, were present from the earliest times. No doubt the seas also swarmed with minute zooplankton feeding off the tinier plants. Such zooplankton were mostly soft bodied and left hardly any record. The carapaces of some extinct crustaceans are known in abundance in graptolitic deposits; they may have fulfilled the same plankton-feeding function as the open sea shrimps today, but there was certainly no early Palaeozoic 'whale' to harvest them. Cherts accompanying ophiolites often contain the remains of microscopic radiolaria, which had acquired their siliceous skeleton even by the Cambrian, and must have had planktonic habits then, as at present.

The graptolites became extinct early in the Devonian, and no ocean-going animals of this colonial type are found subsequently as fossils. At about the same time the early relatives of the ammonites were adapting to life in the open seas, and their coiled shells, variously ornamented, become the fossils most frequently encountered in deeper sea sediments for several hundreds of millions of years. Ammonites swarmed in vast numbers both in the seas of the continental shelves and in the open sea, and the total mass of living matter they must have represented over their geological lifespan from the Devonian to the Cretaceous is almost inconceivable. Some species may have swum in masses, moving gregariously like their distant relatives the squid, which occur in huge numbers in the open seas today. By the later Palaeozoic fierce predators added another level to the food chain, the sharks by then being well advanced, and doubtless as voracious as they are in

left: Clumps of graptolites of only one or two species can be preserved in deep-sea deposits. These Upper Ordivician diplogratids *Orthograptus calcaratus* (450 Ma old) are from Dumfriesshire (Scotland, UK).

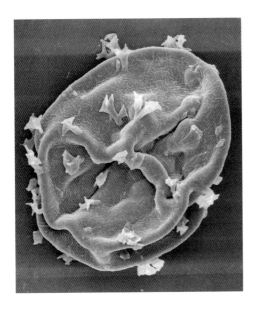

left: *Visbyshaera oligofurcata*, a microfossil belonging to a group of marine phytoplanktonic organisms known as acritarchs that teemed in Silurian seas about (415 Ma ago). From a sample collected near Ludlow (Shropshire, UK).

modern seas. The sharks are an ancient group of fish in their structure and organization, but most of them are superbly well adapted to their role as swimmers and hunters, which has ensured their survival into modern times, while their early prey species, such as ammonites, have passed into extinction. Along with ammonites other kinds of invertebrates, like delicate clams, sometimes occur in profusion in Palaeozoic and Mesozoic rocks. These may either have been free swimming forms, or attached to floating seaweed. The microfossils include an increasing variety of species of planktonic algae which by the Jurassic included forms related to those still living today. The radiolaria continued their unbroken history from the earliest fossil-bearing rocks. In the Jurassic the foraminifera, hitherto almost exclusively bottom-living forms, took to life in the surface waters of the oceans, and played an increasing role as sediment makers, which continued until the present day. During the Cretaceous minute calcareous platelets which coated the outside of algae (coccoliths) were an important component of pelagic sediments, and for all their small size show a wonderful variation in symmetry. The slow build-up in the variety of animals contributing to oceanic sediments continued until the late Cretaceous, when, apparently quite rapidly, the ammonites became extinct.

Whatever the cause of the demise of the ammonites it did not affect the foraminifera or coccoliths to quite the same extent, although the post-Cretaceous foraminifera are all new forms. For Cretaceous and younger

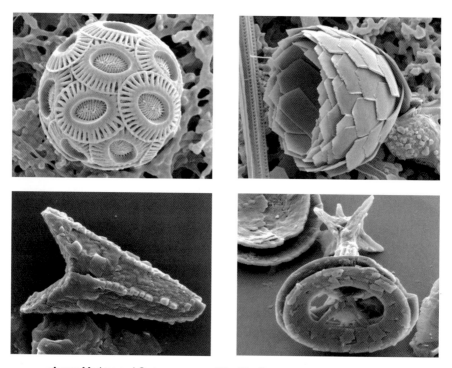

above: Modern and Cretaceous coccoliths. The Cretaceous was the age of chalk due to flourishing of coccolithophorids (microalgae with calcareous exoskeletons). Like planktonic foraminifera they were almost wiped out by the extinction event at the end of the Cretaceous, but a few species survived to repopulate the Tertiary oceans. *Emiliana huxleyi* (top left) and *Florosphaera profunda* (top right) are complete spheres of coccoliths from modern oceans; *Ceratolithoides aculeus* (bottom left) and *Prediscosphaera cretacea* (bottom right) are isolated coccoliths from Cretaceous chalks.

rocks we can investigate the deposits of the deep seas directly by taking cores from the bottom of the sea, or looking at rocks where they have been uplifted on volcanic islands. During the Tertiary the oceans acquired a modern aspect, with fish, whales, modern crustaceans, and micro-organisms related to those of present seas. Nonetheless a succession of different species can be recognized, with certain types becoming extinct, and others appearing as 'landmarks' for the divisions of the last 70 million years.

The sheer volume of the oceans makes the oceanic environment relatively stable compared with the terrestrial one. Extreme changes of temperature, for example, are muffled in the sea, and below a certain depth the temperature is virtually the same the world over. Particularly in the open ocean the animals are cushioned from the effects of environmental fluctuations which have provoked the rapid evolution of terrestrial organisms. Comfort usually stimulates laziness, but we have seen that the oceanic environment has, in fact, changed considerably since the time in which it can be first recognized in the rocks. The radiolaria and planktonic algae have been with us all the time, but have not remained static. Successful groups, like the graptolites, have failed entirely, while the planktonic foraminifera have successfully adapted to the challenges of the open ocean. It is now known that there were certain periods when the oceans did undergo crisis. These times often entailed the increase and spread of sea water lacking oxygen, and were unfavourable to virtually all forms of life. No doubt these crisis periods were implicated in the changes that occurred in the composition of oceanic fauna and flora. However, it can be shown that the changes that occurred on land, for example among the mammals in the early Tertiary, did take place more rapidly than those detectable among the planktonic foraminifera. Nonetheless, the cumulative and inexorable changes have resulted in a complete transformation of oceanic life over 600 million years.

NICHES NOT REFLECTED IN THE ROCKS

The sediment surface indiscriminately receives the scraps that will become fossils, and it would be a mistake to assume that any animal when alive necessarily lived where its remains finish up. Any one environment is usually subdivided into innumerable micro-habitats. The variety of animals living together is explained by the number of niches (particular ways of earning a livelihood) into which a broader environment can be divided. Many of the details of the niche are not reflected directly in the rocks, and therefore the details of the life habits of the fossil plants or animals have to be inferred from different evidence. There are cases where an animal's livelihood depends on numerous kinds of organisms which, being soft

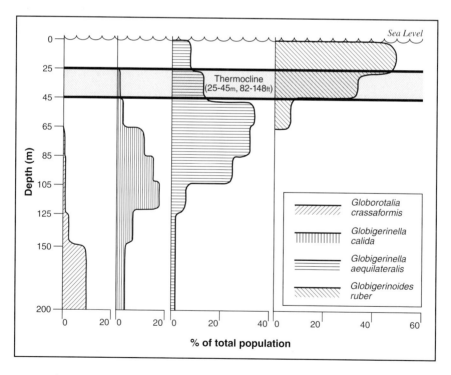

above: Depth stratification to show that different species of foraminifera live at different depths. The sea floor receives dead remains regardless of the depth at which the animal may have originally lived.

bodied, have left virtually no record. Polychaete worms are an abundant source of food in shallow marine environments, but their only geological legacy in most rocks are slight disturbances of the sediment produced by their burrowing activities.

Among living planktonic organisms there is often a stratification according to water depth; most species live near the surface, but there are some that live at deeper levels in the water column. The sediment surface below receives both kinds with equal ease when they die. Depth stratification of this kind has been suggested for fossil foraminifera and graptolites. In this case the record in the rocks can be used to test the theory. Shallow water planktonic species will have a distribution through much of the world in the surface waters, but the deep species will be restricted to areas where the water depth is sufficient. The deepest basins should have the richest assemblages of species in the sediments, because it is in such areas that all the depth zones will be stacked up above the sea floor.

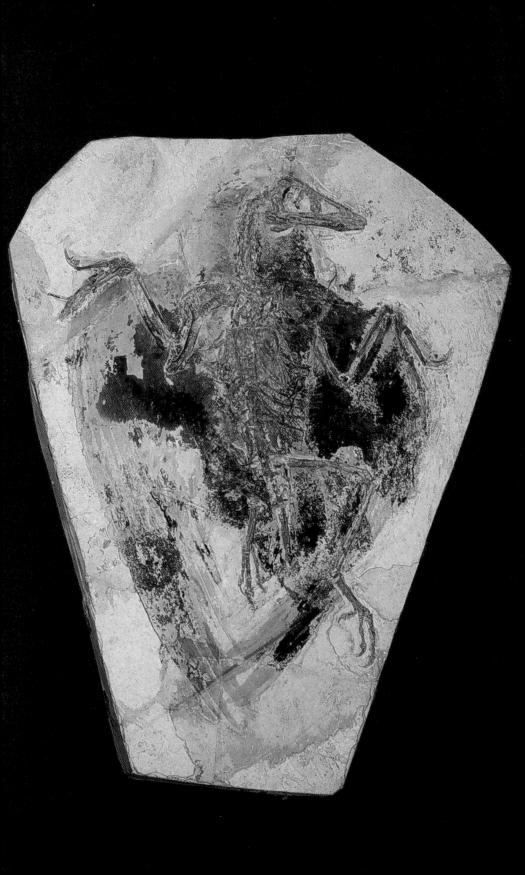

Chapter Four

BRINGING FOSSILS
BACK TO LIFE
. . . .

IN this chapter some of the ways in which palaeontologists determine the way fossil animals lived in the past are described, reanimating the dead fragments to build up a living creature. Since popular ideas of life in the past are often founded on vividly coloured reconstructions of 'The World in the Jurassic' and the like, it is important to remember that these imaginative scenes are all inferences from bones and similar fragments. In fact there is nothing fixed about such interpretations, because the way the fossils are understood may change over the years, and it is usually some time before new discoveries percolate into the popular presentations. Even the most solid-looking dinosaurs may have changed their habits in the last few years!

opposite: *Confuciusornis sanctus*, a 'feathered' dinosaur from Liaoning Province, People's Republic of China; from the early Cretaceous about 124 Ma ago. Males were larger than females and sported a pair of long, narrow tail feathers.

Many different lines of evidence may be used to flesh out the bare bones of the fossils. The first of these is evidence from the rocks from which the remains were recovered. The sediments themselves reveal much about the environment of deposition, as was shown in Chapter 3. It is important to establish whether the fossil animal actually lived in the environment which furnished its sedimentary cover, or whether its remains were swept in from some other place. Fortunately it is usually easy to spot such intruders. The type of sediment and the associated fossils show whether the environment was marine, freshwater or terrestrial, providing the basic information into which the ecology of the animal has to be accommodated. The sediments themselves may have preserved some of the tracks left by the animal to give direct evidence of its past activities. From the tracks alone it is possible to be certain of the bipedal stance of certain dinosaurs, and to measure their stride. The character of the rocks and their setting at the time when they accumulated, provide evidence for the climatic setting in which the extinct fauna lived. Climate imposes certain restraints on possible modes of life; savanna animals differ from those of tropical rain forests, and these again from inhabitants of the tundra at high latitudes.

RECONSTRUCTING FOSSILS

Looking at the fossil animal itself, the first necessity is to reconstruct it as accurately as possible from its fragmentary remains. Sometimes this is a very complex business, particularly for vertebrates with large numbers of small bones. To proceed from the reconstruction to an assessment of probable life habits, two different but complementary approaches are used. One method attempts to compare the structures of teeth or limbs or some other feature of the extinct animal with living analogues. These do not have to be biologically related organisms, the basic argument being that structures that are similar were probably adapted to a similar function. Sometimes these structures are obvious: the ferocious teeth of a predatory dinosaur are a sure indication of hunting habits, with their edges honed into cutting instruments. The structure and function of grinding or chewing teeth in mammals, in which opposing teeth co-operate in action, and which can be matched in extinct, unrelated mammals, is a much more subtle matter, involving detailed studies on the operation of living dental systems to help elucidate the functioning of fossil ones.

The technique of 'hunt the analogue' is a favourite one practised by palaeontologists, but is certainly not foolproof, because there are many fossil animals that defy comparison with living organisms, and some analogues do not stand up to detailed scrutiny. The second method tries to analyse the structure of the fossil. If the fossil is constructed in a certain way, then there are only a limited number of jobs that the structure could perform. The idea here is to decide which of the possibilities is the most likely. This assumes that nature only manufactures efficient designs, and for the most part this is a reasonable assumption. Most animals today do seem to have bodies that accord well with the functions they have to carry out to survive: flyers are aerodynamically efficient, active swimmers have suitable streamlining, herbivorous mammals have teeth appropriate for grinding plant food. Ideally one could construct a model of the fossil to test out these various functions, but the number of examples where the analysis has been pursued this far are limited. Of course it can never be known whether the right answer has been reached (unless somebody dredges up a 'living fossil'); there are only varying degrees of probability. The best answers are obtained where the functional design of the fossil points to the same life habits as a living analogue, and where both are consistent with the geological circumstances in which the fossil remains occur.

SWIMMING TRILOBITES

There are thousands of different kinds of trilobites. All of them are marine and all of them are extinct. Since many kinds of trilobites coexisted at any one time they were probably occupying different ecological niches, behaving in different ways, corresponding with the wide variation in shape that they show. We can use the methods described above to elucidate some of these occupations, and get a glimpse, albeit an imperfect one, of the trilobite as it lived. The following example is one where both methods can be used, and where all the evidence points in the same direction, so that the answer is probably correct.

Most trilobites are oval shaped being rather longer than wide, not greatly convex, and with eyes that occupy perhaps a quarter of the length of the head. There are a few trilobite species, however, with enormous, globular eyes. By examining the way these animals are put together it is possible to suggest a likely mode of life.

We can take the eyes first. The trilobite eye is a compound type, and each lens is made up of a calcite crystal. From the optical properties of this mineral we know that the lens is able to interpret light coming from a direction more or less at right angles to the lens surface. We can therefore deduce what the field of vision of our giant-eyed trilobite was, by looking at the directions in which all the lenses face. It turns out that our animal was able to see in almost every direction — upwards, downwards, sideways, forwards, and even backwards, because the eyes bulge out beyond the line of the rest of the body. Most other trilobites have a predominantly lateral field of view. If the animal lived on the sea bottom, it seems unlikely that it

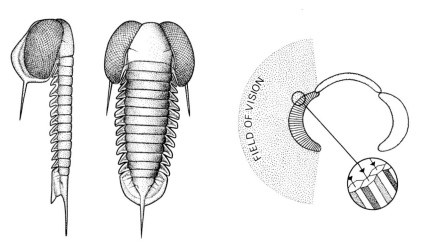

left: Giant-eyed Ordovician trilobite, seen from the top and side (twice its natural size).
right: Section through the lenses of a giant-eyed Ordovician trilobite to show the direction of the field of view.

would have had eye lenses specialized for looking downwards, and so we begin to suspect that the animal habitually dwelt above the sea floor. Other features of its shape are consistent with this. At the edge of the eyes is a pair of long spines, and these point downwards at a steep angle, at a very awkward attitude for resting comfortably on the sea floor. Most 'normal' trilobites have a more or less horizontal rim around the forward margin, which may have rested on the sediment surface.

The thorax of the giant-eyed trilobite is remarkably long compared with the average trilobite. The convex, middle part of the thorax contains the musculature that operates the appendages (which, as usual, are not preserved), and the relatively large volume of this region shows that the musculature was powerful. The construction of this giant-eyed trilobite suggests that it lived an active swimming life well above the sea floor, and possibly in the surface waters of the sea. Because there are no living trilobites we cannot find direct confirmation of this hypothesis, but we can look for other arthropods in the present day oceans that have the same modifications of the eyes. The analogy is found in the deep sea crustacean *Cystisoma*, which also has enormously expanded eyes, looking like headlamps, compared with its bottom-dwelling relatives. In fact many of the arthropods that inhabit the water column have large, globular eyes of this kind.

above: Giant-eyed living deep sea crustacean (*Cystisoma*).

If it is correct that our trilobite lived above the sea bottom, actively swimming in the water, there are certain predictions we can make about its geological occurrence which can be tested by looking in the rocks. Most bottom-dwelling trilobites preferred to live at a particular water depth, or on a particular type of sea bottom (mud, sand or lime). No such restriction should apply to our globular-eyed species; it should be found along with all other different kinds of trilobite assemblages without preference. This has been proved in several places: in some Arctic localities it is found with trilobites that lived at great depths in the muds of the Ordovician ocean, while in Canada the same species occurs mixed with the inhabitants of the shallow water seas, where limestones were accumulating. By the same token the free-swimming trilobite may be expected to have a very wide geographical distribution, for oceans would provide no barrier. Again this proves to be the case: our example is found in Arctic Canada, in the Arctic island of Spitsbergen, in the deserts of Nevada and Utah in the United States, in western Ireland, in Russia, and in north-west Australia.

All the lines of evidence described above suggest that these giant-eyed trilobites of the Ordovician were active swimmers in the surface waters of the oceans, detailed consideration of the way the trilobite is constructed, analogy with living animals with similar adaptations, evidence from the rocks and the distribution of the fossils.

There are a number of questions we can ask about the life habits of these trilobites which are not subject to such careful scrutiny, for example what did they eat? It is not possible to see directly how the mouth appendages functioned in feeding, and the stomach contents are not preserved. So here we can only speculate. Active swimmers in the surface waters of the present oceans are likely to feed directly on plankton, and the trilobites may have had a method of harvesting large quantities of such food. Alternatively they may have been hunters of larger prey, in which case when the appendages are eventually discovered they may prove to have adaptations for grasping and manipulating larger food. Puzzles remain, even though we can be confident of the rudiments of the story.

CONTROVERSY AMONG THE GIANT DINOSAURS

The giant sauropod dinosaurs like *Brachiosaurus* were the largest land animals the world has seen. They weighed more than 80 tonnes and present quite different problems in the interpretation of their life habits from the diminutive trilobites. Such spectacular animals have obviously attracted much attention, and one might expect that the problem of how they lived would have been satisfactorily solved long ago. Many old-fashioned popular books showing the Mesozoic giants in their natural setting portrayed them wallowing about in swamps flanked by deep vegetation, their bodies largely under water. Surely an animal of this bulk, it was argued then, must be partly supported by water, and their relatively inconsequential teeth must have been adapted for chewing on the kind of soft, luxuriant vegetation that flourishes in and around swamps. Because

above: One set of dinosaur trackways seems to show *Apatosaurus*, a sauropod dinosaur, walking on its front feet only! Scientists have concluded that it was floating in water, pushing itself along with its front feet and steering with its back legs.

their nasal openings were on the top of their heads they could even continue to breathe if it became necessary to submerge totally.

More recently the life habits of these giants have been looked at in a way that disproves most of these erstwhile notions. Consider the structure of their legs. The sauropods have relatively long, pillar-like legs, resembling those of the elephant, the largest living land animal, and may have been well adapted for supporting the huge bulk of the animal. The feet of the sauropod are small (relatively speaking), with short, stubby toes, yet animals that walk on soft mud tend to have spreading feet to distribute their weight more evenly. It is difficult to see how the compact feet of the sauropod could avoid becoming stuck fast in the soft, muddy bottom of a swamp. If the dinosaur did, after all, live on dry land, then the long neck could have usefully functioned to allow the animal to browse the high foliage of trees. This is how the animals are portrayed in recent films such as *Jurassic Park* and *Walking with Dinosaurs*. Their fossil remains occur with other animals and plants, which are generally accepted as being terrestrial.

above: Footprints and trackways tell us whether dinosaurs lived alone or in groups, and how fast they moved.

above: Four-legged animals with a backbone walk in one of three ways. Monitor lizard (left): sprawling, with legs at right angles to the body; the early ancestors of dinosaurs walked like this. *Euparkeria* (centre): legs straighter and body held high off the ground; this can only be maintained over short distances. *Triceratops* (right): straight legs tucked under the body; the upright stance was the key to the dinosaurs' success.

There is some evidence from the tracks they have left behind that *Diplodocus* and its allies moved about in herds. They may have been the gigantic reptilian analogue of the elephant, and it may be no coincidence that the elephant also has its nasal openings on top of the skull, with the nostrils in this case sited at the end of the trunk — it has even been suggested that some sauropods may have had a proboscis of some sort.

The balance of evidence overwhelmingly supports fully terrestrial habits for these giants. Other aspects of the dinosaur living habits have been the subject of debate. There was much controversy as to whether the dinosaurs as a whole were cold-blooded, like all living reptiles (and there is no doubt that dinosaurs were reptiles), or warm-blooded, resembling mammals. Cold blooded animals have to 'warm up' before they can be fully active; that is why lizards and snakes bask in the Sun in temperate climates. For this reason they cannot cope with climates having greatly extended winters. Warm-blooded animals have the same body temperature at all times, and can be continuously active, but they use far more energy and hence need more food than cold-blooded animals of the same size. The posture of many dinosaurs, particularly the carnivorous theropod dinosaurs, was fully erect with the legs beneath the body, unlike the sprawling legs of living reptiles. The long back legs of such hunters look highly suitable for running, and as they did so the long tail may have been held erect as a kind of counter-balance. For any kind of prolonged activity, warm-bloodedness would have been a distinct advantage. There was, until recently, a vigorous argument between the cold-blooded and warm-blooded schools. Detailed study of the structure of the bones of dinosaurs has found favour in both schools of thought. When they were young, it is likely that dinosaurs grew very fast, and their bone structure is consistent with a kind of warm-bloodedness like that of mammals today. However, as they grew large, growth slowed down dramatically, and their huge size allowed for a considerable degree of temperature control. They managed to have the best of both worlds, combining reptile energy efficiency with mammalian levels of activity.

The warm-blooded theorists allied themselves at one time with the theory that the dinosaurs included the ancestor of living birds. The Jurassic bird *Archaeopteryx* is one of the most famous fossils, and was a contemporary of the dinosaurs. Some of the smaller, highly active dinosaurs were about the size of a chicken, and there is more than a passing similarity between a running ostrich and the kind of reconstruction that shows fleet-footed, running dinosaurs. The bird-dinosaur connection got a tremendous boost at the end of the 20th century with the discovery of a whole series of 'feathered dinosaurs' which proved a connection that anatomists had previously inferred from the study of bones alone. Feathers are, of course, otherwise the exclusive property of birds. These important fossils were discovered in China, and there has been a race to describe the most spectacular example — one was even exposed as a fake. Few scientists today question the bird-dinosaur connection. Most specialists regard the

 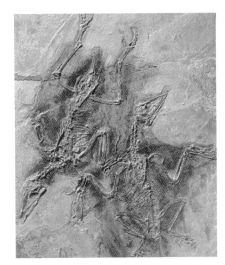

left: *Archaeopteryx*, the earliest known bird (147 Ma old), found in Germany. It shows a reptile-like bony tail and claws on the fingers, as well as feathers and a 'wishbone' which are characteristic of birds.
right: Two 'feathered dinosaurs' *Confuciusornis sanctus* (124 Ma old) from Liaoning Province, P.R.C. Thousands of individuals recovered from this site suggest that they might have lived in large colonies.

dinosaurs and the early birds as having similar specialized metabolism. Truly warm-blooded birds may have arisen as they mastered the difficulties of the aerial habitat. Possibly they retained the same high metabolic rate as has been ascribed to young dinosaurs.

GRAPTOLITES – FLOATING COLONIES

Around 1830 geologists were beginning to unravel the mysteries of the early Palaeozoic rocks. The past world recorded in the rocks could not have seemed more alien. Many of the organisms that left fossil remains were now extinct, although some, like the trilobites, could obviously be placed into a phylum with many living representatives. The graptolites, however, were initially completely enigmatic; they were first described as plants! Their remains were so abundant that they could not be ignored. Their colonies, looking like miniature hacksaw blades, often completely covered bedding planes, and were usually found in the absence of other kinds of fossils. It became apparent that they could be useful in subdividing the intractable stretch of time from the late Cambrian to the Silurian as they changed in obvious ways from one rock formation to the next. Graptolites with numerous branches seemed to dominate the earlier rocks, ones with fewer branches appeared later, while in rocks we would now recognize as Silurian and early Devonian, forms with a single branch (or stipe) were abundant. With the discovery of better preserved material it became apparent that the graptolites consisted of rows of tiny cups which were interconnected by a common canal and that they were colonial animals. We now know that the graptolites are an extinct branch of the pterobranch hemichordates, an insignificant group today, consisting of a few encrusting colonial organisms.

How did these mysterious organisms live? In this case it is difficult to apply the analogy method, because the graptolites are different from any animals now alive. The modern pterobranch hemichordates are all encrusters, and virtually all colonial organisms (such as corals or bryozoans) are bottom dwellers in the sea. From this argument the graptolites should have been bottom-dwelling colonial organisms, living by filtering small particles of food from the water. Some of the graptolites (dendroid graptolites) were shrubby colonies,

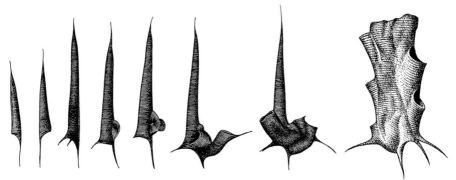

above: Growth and budding of an Ordovician graptolite colony, enlarged about 20 times.

looking very much like living hydroids or some bryozoans. These were equipped with roots, and an inference of bottom-living habits seems reasonable for them. The great majority of the class, however, including all those with typical saw-blade profiles (graptoloids) which are widely used in the dating of rocks, lacked any kind of rooting structure. When they were looked at in more detail it was found that the colonies grew from one single tube (sicula) which is often facing in a different direction from the tubes inhabited by the rest of the colony. A slender rod (nema) usually grows from the top of the sicula. Since some of the graptoloids are large and robust, this slender rod served an inadequate basis for attachment to the sea floor, particularly in any sort of turbulent environment. Then there were some forms, like *Phyllograptus*, in which the colony grew over the slender nema to leave no visible means of attachment at all. It begins to look as if the graptoloids could have been free-floating animals, of a kind without any living counterpart.

above:
Phyllograptus, a graptolite colony with no means of attachment (twice its natural size).

The geological circumstances in which the most abundant graptolite faunas are found can now be introduced into the argument. The most typical occurrence of these fossils is in dark, often sooty, black shales. In many localities graptolites are the only fossils to be found, but they usually occur in abundance. Sometimes they occur within cherty rocks that contain the remains of planktonic radiolaria. Some of the graptolitic black shales are what we can recognize, with the hindsight of plate tectonic theory, as the deposits laid down in a truly oceanic environment. These are frequently associated with volcanic rocks of oceanic type, so there seems little room for doubt that the graptolites were capable of living in an open ocean environment. The fact that they occur in shales without other benthic remains makes it very likely that they were free-floating, and this is consistent with the way some of the species were constructed. Dead graptoloids simply drifted down to the bottom where they were preserved by dark muds at a depth at which there was little oxygen in the water to support bottom-dwelling organisms. A few examples have been found where a number of graptoloids seem to have been associated together, attached (by the nema) to a 'float'. Other species appear to have the nema extended into a kind of vane. There may have been some species which were attached to floating seaweed. Many graptoloids would have drifted into shallower water, where they are associated with a shallow marine fossil assemblage. Like the pelagic trilobites, individual graptoloid species are very widespread, which is what one would expect of an animal with the wide ocean as its habitat. It is not surprising to find that this widespread species have been used extensively to correlate rocks between different continents — they form the standard for the Silurian period, where the graptolite faunas are nearly the same all over the world.

In the case of the graptolites a combination of geological evidence and the shape of the animal itself has been used to say something about how the animal lived. Many problems remain which we are only just beginning to understand. Which way up did the graptolites float? Which species were adapted for living at particular depths? Why did graptolites have an evolutionary trend from many branches to few branches? There are plenty of different answers available to these questions, but, like the metabolism of dinosaurs, no consensus has been reached as yet.

ANIMALS WITH TWO VALVES
Graptolites, trilobites and dinosaurs are all extinct, and so in the examples we have considered so far there is no direct recourse to the study of living animals to help us bring our fossils back to life. Any analogies have to be

with living creatures that are distantly related in the zoological sense, although there may be reasons to suspect that they have similar life habits to those that the fossil animals once enjoyed. Some of the fossils with the simplest construction are those with the soft parts encased within two valves, principally the bivalves and the brachiopods. These two groups of animals are unrelated, but both have living representatives in the oceans, and so it is possible to look at living animals to help with the interpretation of fossil examples.

Of the two, the bivalve molluscs are much more significant in the sea today. The brachiopods are still numerous in some environments, but these tend to be rather inaccessible, such as in the Antarctic, which inhibits their direct study in the field. Most brachiopods lead a rather uneventful life, attached by a stalk to a hard surface, like the underside of a rock, or another shell, filtering food from the surrounding sea water. There is nothing to suggest that the innumerable fossil brachiopods had any other method of feeding. There are a great variety of shapes among the fossil forms, many of them unmatched in living species. There must have been many different ways for brachiopods to exploit their simple mode of life.

For a filter-feeding animal, it is sensible to separate the inhalent current, bringing food, from the exhalent one, which carries away waste products. Brachiopod feeding is carried out by the lophophore, a ciliated band usually carried on a loop, which also creates the currents used in feeding. Many brachiopods have a depression in the wavy margin of the valves with the middle bent downwards. This form of the shell reflects the separation of the inhalent currents from the exhalent currents. Currents pass in through the sides of the shell, over the ciliated lophophore where the food is extracted, and then out through the depression in the margin of the valves. The lophophore is supported in several different ways. In some species the support is provided by a spiral made of calcium carbonate. This increases

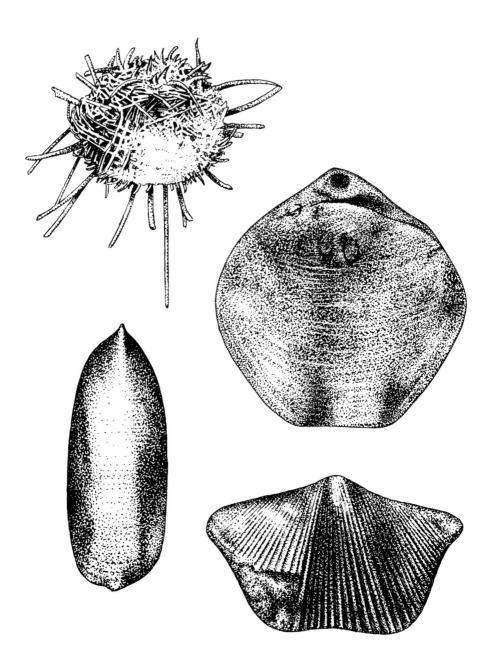

above: A sample of the different shapes achieved by the brachiopods; different shapes indicate different modes of life.

the length of the lophophore available to extract food from the inhalent current, and so increases the efficiency of the process. Brachiopods are rather more complex animals than their simple shape suggests. Many species have a finely folded valve margin, crumpled like corrugated cardboard. Since the shell grows at the margins, a ribbed shell is produced, with fine or coarse ribs according to the species. A ribbed shell of this kind is often stronger than one without ribs, and it is probably no coincidence that ribs may be developed on bivalves too. The wavy margin may also serve another function as it increases the length of the margin, allowing the animal to admit more food per unit of length. It also prevents the entry of irritating particles of sand, with which the animal cannot cope. However, there is still a problem at the crests of the corrugations, where the gap between the valves is larger; many brachiopods solve this particular problem by having fine, hair-like spines which cover the crests of the corrugations, and so prevent the intrusion of unwanted particles.

left: Spiral supports for the brachiopod feeding system preserved inside one of the valves (etched out with acid).

Although most brachiopods were attached to a firm substrate by a stalk which emerges through a hole in the pointed end of the animal, there are some brachiopods in the fossil record in which there is no opening for such a stalk. These must have lain freely on the surface of the sediment. In some of these species one valve is greatly thickened. This is likely to have been the lower valve, and fossils are sometimes found in life position which confirms this. The other valve sits on the thick one like a lid, and the whole animal is curved so that the margins of the valves are kept clear of the sediment. The end result is an animal that looks very much like some kinds of oyster, although a glance at the internal feeding structures shows at once that they are brachiopods, unrelated to the bivalves they superficially resemble. Some species have long spines attached to the lower, thicker valve which could possibly have acted as anchors, securing the animal in the sediment.

Unlike the brachiopods, many of the modes of life of living bivalve molluscs can be more or less directly matched in fossil examples. By studying how the life habits of living bivalves are reflected in the shapes of the shells we can make deductions about how the fossils lived. One major difference between brachiopods and bivalves is that the latter are capable of moving freely, using their foot to crawl or dig. Some bivalves, like mussels and oysters, do remain fixed throughout their adult life, and it is possible that these fill the ecological role today of some of the large brachiopods of the Palaeozoic. Some of these lived in shallow water sites and were gregarious like mussels. However, since brachiopods were unable to do many of the things that bivalves do very efficiently, it would be unwise to attribute the decline of the brachiopods to diversification of the bivalves.

Burrowing bivalves, which are typically found in the soft bottoms of shallow seas today, dig themselves down into the sediment, where they escape the attention of many predators (except birds with long bills).

Many of them retain contact with the surface of the sediment by means of long siphons, which enable them to breathe and feed on small organic particles. Most bivalves with siphons develop a gape at one end of the valves, so that they do not close entirely in the region from which the siphons protrude. It is easy to recognize such gaping valves in fossils, and there are accompanying changes in the internal structure of the shell which are associated with species with long siphons. Hence bivalves from Palaeozoic times can be identified with some confidence as being ancient burrowers. This has been confirmed by discoveries of fossils preserved in their life position. Often the process of 'digging in' is assisted by a characteristic pattern of chevron shaped ribs on the surface of the shell, which are also obvious features of some fossil shells. Such V-shaped ribs are not found on brachiopods.

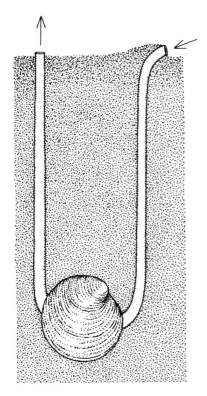

below: Burrowing bivalve mollusc in life position; the siphons retain contact with the surface.

Other kinds of bivalves have used their capacity for free movement to become efficient swimmers, clapping their valves together like castanets to move through the sea to find new feeding grounds or to escape the threat of predators. These bivalves have fan-like shells, often strongly ribbed (*Chlamys*). While the bottom-dwelling or burrowing bivalves mostly have a pair of strong muscles to pull the valves together, in the swimming forms the muscles have become modified so that there is one particularly powerful muscle, centrally placed, to produce the powerful clapping movement that propels the animal through the water. This muscle leaves a conspicuous scar on the inside of the shell, an impression which is easily preserved in the fossil state. It is therefore possible to deduce swimming habits in fossil bivalves by comparison with the living forms. These swimming habits go back into the Palaeozoic; to the Carboniferous or even earlier. There are a few bivalves which are found in rocks deposited in the same sort of black shale environment that was mentioned above in the discussion of graptolite life habits. Some experts maintain that these molluscs were able to swim in the open ocean, or that they were attached to floating seaweed; some such mechanism has to be invoked to explain how these bivalves came into an environment that lacked bottom-living animals.

In the selected examples discussed above showing how modes of life can be deduced from fossils with two valves, the most important point is that relatively complicated life habits can be inferred from a careful consideration of the anatomy of even the simplest looking animal, particularly when there are related, similarly adapted animals alive today for comparison.

PALAEONTOLOGICAL ENIGMAS
Every now and then the fossil record throws up fossils which are palaeontological puzzles. They are obviously the remains of some kind of

animal, but the problem is to decide what kind. They tend to be rather rare and preserved in a special way. Like so many palaeontological matters, they stir up arguments between specialists who think they have a way of solving the enigmas.

Some of these puzzling fossils are quite small. A few years ago a minute fossil only about 2 mm long was recovered from limestones of Ordovician age, and christened *Janospira*. It looks remarkably like a trumpet. The 'mouthpiece' and the 'horn' of the trumpet are both open, and a coil hangs down from the middle. The end of the coil is closed, and it seems reasonable to assume that the animal started growing as a coiled shell. It must have then changed its mind, and started to grow in two directions, a narrower tube into the mouthpiece and a broader one into the horn, and there is some evidence that the horn end continued to get wider and longer. The problem is that it is hard to equate this kind of growth with that of any known animal group. It looks like some kind of mollusc, but no mollusc fits easily into this pattern of growth; some have suggested it might be some kind of snail, or perhaps a monoplacophoran. It remains a puzzle. It is found along with fossils of planktonic organisms, and it is possible to explain the change in growth between coil and trumpet as a change that happened when the larval shell settled on the sea bottom. But since virtually all phyla of animals are found with planktonic larvae, this is no help in solving the puzzle.

Some larger fossils are even more puzzling. A number of fossils from the famous mid-Cambrian Burgess Shale of western Canada are so strange that they have even been claimed as the representatives of 'extinct phyla'. One of these, *Hallucigenia*, is an animal with what is thought to be a gut with a set of tubules arising from it, and below this are curious pointed structures. A suggestion was made that early on in metazoan evolution there were a number of experimental designs which did not give rise to direct

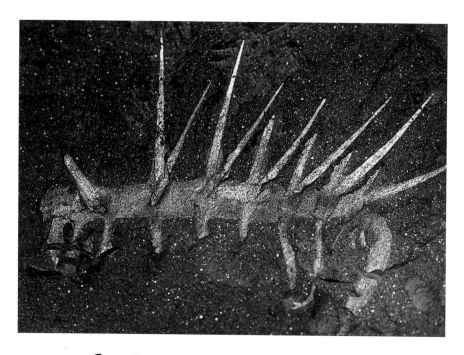

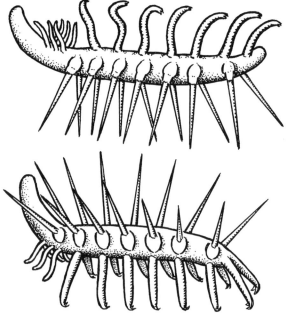

above:
Hallucigenia sparsa
left: Older and
bottom left:
more recent
reconstructions
of *Hallucigenia*
turning it
upside down.

descendants; naturally these do not fit into the pigeon holes based on living animals. *Hallucigenia* was claimed as one of these. However, subsequent studies of this oddity have shown that the early interpretation was completely wrong. It was upside down! Further investigation revealed that the tubes on the 'back' were actually legs, and the pointed structures were spikes on the back of the animal. *Hallucigenia* is related (distantly) to the velvet worm *Peripatus*. It is still a very odd creature, but its place in nature is now explicable. One of the fascinating things about palaeontology is that new explanations of such puzzles cast new light on the history of life.

Chapter Five

ORIGIN OF LIFE AND ITS
EARLY HISTORY
. . . .

WHEN the Earth formed about 4500 million years ago there was no life. By about 3500 million years ago the first, tentative traces of life are to be found in the rocks. The profound series of changes that have occurred since the Cambrian, and which are the subject of most of this book, took only a fraction of time in relation to the history of the Earth. The time of the origin of life is very inaccessible to direct study, all certainties about the nature of the atmosphere and configuration of land and sea are gone, and speculation has more room for manoeuvre with this palaeontological mystery than with any other. What is certain is that the Earth's surface at the time of the origin of life must have been different in almost every respect from that of the present.

Left: The Western Hemisphere of the Earth's surface in 1997, created from the data of three different Earth observing satellite instruments. At the time of the origin of life the Earth's surface would have been very different to what is shown here.

THE EARTH AND ITS ATMOSPHERE

The formation of the Earth was but one incident in the formation of the solar system as a whole. If present ideas are correct, the Earth accreted from small particles swirling in a disc-shaped nebular cloud. The planet grew from a 'seed' of magnetic metallic grains, which attracted more material by gravity, so that it grew rather like a snowball. Early on, the Earth melted and a molten core was formed, largely composed of the metals iron and nickel. Accretion of other material continued, with the addition of volatile components like water, carbon dioxide and chlorine when the body was large enough to retain them. It is unlikely that the Earth retained much of an atmosphere at this time. Then the surface of the Earth cooled quickly to form a thin, brittle crust. Massive bombardment with meteorites continued; the surface of the Moon preserves this stage in the evolution of the Earth as it is covered with impact craters. The moon itself may have been the result of a massive impact on the 'proto Earth'. The Earth incorporated the additional meteoritic material into the upper part of its mass. The gravity of the Earth was powerful enough to hold on to some of the more volatile ingredients, particularly water, without which there would have been no life. An atmosphere gradually formed around the Earth, and the incessant bombardment by meteorites slowly abated. The rate of heat flow from the Earth's interior was high at this early stage, and volcanic activity was widespread and continuous. Any volcanic eruption is accompanied by the release of huge quantities of steam and gases like nitrogen and carbon monoxide, as well as highly toxic acids like hydrogen chloride and sulphur dioxide. We can visualize acid rains attacking rocks and reacting with them to form minerals such as sea salt (sodium chloride). The steam released into the atmosphere from the volcanoes condensed to form seas, and since the first traces of sedimentary rocks are as old as 3800 Ma or more, there was enough water for sedimentation by this time. Presumably the processes of erosion had also begun to shape out patterns of cliffs and beaches so that the first vestiges of the modern topography appeared.

In several important respects the environment of the early Earth was completely different from today. The original atmosphere may have had a much larger proportion of carbon dioxide than is present in the current atmosphere. There was little or no free oxygen in the atmosphere. This meant that there was no ozone layer (a cloak of modified oxygen high in the atmosphere today that acts as a screen to prevent harmful radiation getting through to the Earth's surface). The penetration of such light to the surface of the Earth produced the kind of chemical reactions that might have led to the necessary building blocks for life itself, including the splitting of the water molecule into its component atoms. These chemical reactions can be reproduced in the laboratory — the right mixture of chemicals and gases, charged with the effects of ultraviolet light — and sure enough some of the basic ingredients of life appear after a while.

EARLY LIFE

The presence of amino acids in the early Earth was particularly important, for chains of such acid molecules join together in their hundreds to produce giant molecules, forming the basic structure of proteins, without which there would be no life. All organisms are able to produce copies of

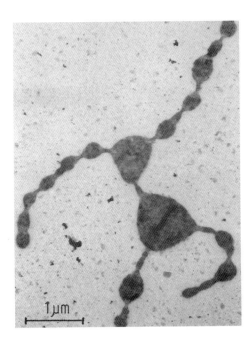

left: *Hyperthermophilic* Archaea which lives in hydrothermal vents at temperatures around 100°C (212°F).

themselves, and presumably it was necessary to have evolved the long spiral chain molecule DNA (or possibly RNA first), which is the basis of the copying process, before the first self-replicating cells could exist. The first living cells were of the simplest kind, represented today by bacteria (especially *Archaea*). Many primitive living bacteria have an 'economy' built on exploiting sulphur compounds, which were present in abundance around the hot springs, fumaroles and volcanoes in the Precambrian. They are tolerant of high temperatures — in fact they need heat to survive. Studies on the genes of such organisms show that they are likely to be the most primitive organisms alive on Earth today. Life probably started at high temperatures, in the absence of oxygen. Whether the vital steps occurred in the early seas or in some other location, such as shallow pools or the hot vents of submarine volcanoes, is still a matter of debate. But presumably once the first true, self-replicating cell had appeared it could spread and prosper unhindered wherever the right conditions for its nourishment were to be found. That Earth, no matter how inhospitable it would seem to us, would have been a Garden of Eden for the first organism. Most living things have so many molecular features in common that it seems feasible that the generation of life happened just once. We are only now beginning to appreciate the huge variety among the bacteria. In some respects they are the most successful organisms on Earth, for they are found everywhere from the Arctic to deep beneath the surface of the Earth in the rocks themselves.

The origin of life requires principally time, the right conditions, and a few special events to link all the components into the finished cell. Recently another idea has claimed a lot of attention, although it is not a new one. Life may have been of extraterrestrial origin. The early Earth was almost certainly bombarded with particular kinds of meteorites (known as carbonaceous chondrites) and these meteorites when they appear at the present time do contain organic kinds of molecules.

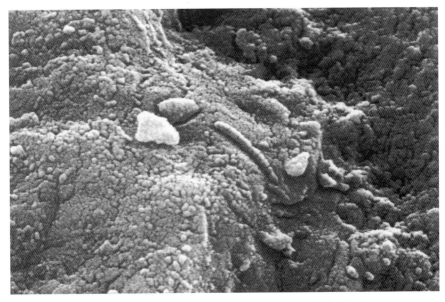

above: SEM image of the surface of the carbonate patches in the Martian meteorite ALH 84001, showing a tube-like 'microfossil'. The organic nature of these microfossils is still disputed.

As far as the origin of life on Earth is concerned it is true that it is possible to produce carbon compounds like those in meteorites under the right experimental conditions, but there is a curious difference between these compounds and those which predominate on Earth. Many carbon compounds can exist in two forms, which are chemically identical, but which have the property of rotating light either to the left or to the right. On Earth there is a predominance of the former which rotate light to the left; but in meteorites the left and right-handed forms of the same compounds are equally abundant. It therefore appears as if organisms on Earth manufacture carbon compounds which are subtly different from those produced in meteorites. It seems preferable to regard the Earth as the cradle of life, if only because this theory requires fewer coincidences to produce an organism adapted to terrestrial conditions, and there has been a vast stretch of time available for the shuffling and rearrangements of chemicals necessary to make the first living cell. Therefore, although meteorites contributed organic molecules to the 'soup' from which life was brewed, it is probable that the cauldron itself was on the Earth and not in the far reaches of space.

Some scientists have claimed that the periodic appearance of extraterrestrial materials on Earth, especially from comets, has continually influenced the evolution of life — not just its earliest history. For example, they suggest extraterrestrial biological material provides an explanation for catastrophic viral epidemics. Such meteorite induced epidemics are one of a number of theories about extinction which it is impossible to test one way or the other. It is difficult to see how meteorites could have been so selective as to remove, say, dinosaurs but not birds (which may be close relatives of the dinosaurs), or ammonites but not snails at the end of the Cretaceous. Comets passing close to the Earth from time to time may well be implicated in some of the crises in the history of life, but this is more likely to have been by secondary effects rather than by direct biological influence. Viruses are rather specialized organisms, often adapted to a particular host, so a virus with fatal properties arriving from 'out there' might be expected to remove only its host species. This does not explain mass extinctions as we actually observe them in the rocks.

One way or another the crucial steps were taken, and cells evolved. It is only in the last two decades that proof of the existence of the oldest kinds of organisms has been recovered from the rocks. As one might expect these rocks are rather special: they have survived from the remotest parts of the Precambrian hardly altered by heat or pressure. They are often cherts, fine-grained hard rocks composed of silica, and when cut into thin slices and ground they become transparent and reveal tiny fossils. Most of the cherts which have yielded fossils seem to have been deposited in shallow water around the protocontinents of the time.

PRECAMBRIAN ORGANISMS

The earliest and most primitive fossils are those of prokaryotes, simple rods, spheres or filaments in which the cell contents lack a defined nucleus (the package within the cell which houses the DNA in all higher organisms). Prokaryotes would be referred to as bacteria today. Such simple prokaryotes are known as fossils in rocks as old as 3500 Ma. (Fossils of eukaryotes with a defined nucleus are known from rocks about 2100 million years old). Even for those used to thinking in terms of geological time these figures are hard to comprehend. The time during which these simplest of organisms held sway is so much longer than that which has elapsed since the beginning of the Cambrian, which used to be thought of as comprising the whole of the fossil record. Momentous events at the cellular level happened during this vast time. The earliest fossils discovered

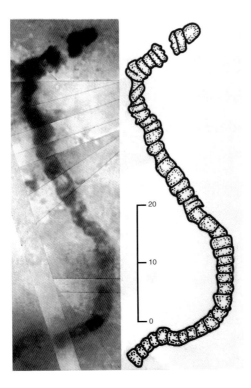

left: Electron micrograph and reconstruction of the fossil claimed to be the oldest on Earth the bacterium *Primaevilium amoenum* from north-western Australia

from cherts can scarcely represent the whole story, new discoveries are still being made, and it is clear that there was an increase in the variety of these small fossils through the Precambrian, even from the patchy record they have left.

Ancient bacteria were able to live in conditions which would be inimical to most living organisms. Some had a metabolism based on sulphur, and these survive today in hot springs and similar habitats. Others acquired the ability to use the Sun's energy to produce nutrients (the process of photosynthesis), but not all of these gave off oxygen, as do the higher plants. Cyanobacteria which did exhale oxygen were certainly present 2700 Ma ago, and some scientists claim that they may have been present 3500 Ma ago.

Some Precambrian fossils had been known about for many years before it was concluded that they really were organic. These include finely layered rocks, sometimes with the external appearance of numbers of small cushions, or occasionally steeper pillars. It was the regularity of their layered structure that suggested they might be of organic origin: such fossils were christened with names like *Cryptozoon* ('hidden animal') when they were discovered in Precambrian rocks from Canada and elsewhere. At the time it was considered rather unlikely that fossils could occur in rocks as old as this. Nowadays, fossils of this kind are known to be the remains of structures produced by cyanobacteria, a perfectly reliable indication of biological activity. They are known as stromatolites. These stromatolites are now known from rocks younger than Precambrian age as well. They are common in limestone rocks originally deposited in shallow water environments in the tropics, and they appear to be especially characteristic of what were quiet water sites between high and low tides. A number of years ago living stromatolite mounds were discovered in Shark Bay, Western Australia, and like some of their fossil counterparts they were

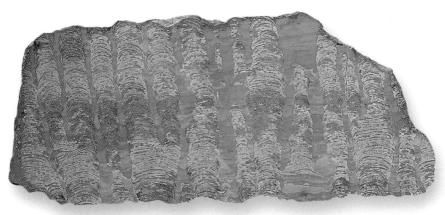

top: Fossilised bacterial/algal stromatolites around 15 Ma old, Wadi Kharaza, northern Egypt.

bottom: Fossil stromatolite cushions from the Precambrian rocks of eastern Siberia (half size).

found in an intertidal environment. Since the oldest of the fossil stromatolites were at least 3000 Ma old, these rather unspectacular mounds take all prizes for the most enduring of living fossils. Why they had not been recognized as of organic origin previously is because the very primitive organisms that make the mounds are not usually preserved as fossils — their simple threads generally decay without trace. The mounds are produced by a skin of tacky cyanobacteria trapping sediment season after season to produce the fine layering characteristic of the stromatolite.

Beneath the surface layer other, ancient kinds of bacteria thrive, so that the stromatolite is really a very simple community, one which has survived from the earliest days of the Earth. Tidal waters drain off through the channels between the mounds. In the living examples there is a certain variation in the form of the mounds according to where they are growing. Some of the same variation has been observed in the fossil forms. Nonetheless, a growing number of palaeontologists studying the Precambrian mounds believe that there is also a variation through time in

above: Living algal stromatolites on the sea floor, *Exuma Lays*, Bahamas.

the shapes of the mounds, so that they can be used to characterize very broad segments of the Precambrian. Their occurrence in Precambrian rocks which were deposited over the epicontinental seas of that time is almost ubiquitous; they have been found over huge areas of Russia, Australia, Greenland, Africa, Canada, the United States and Scotland. They were much more widespread in these far-off times than they are today and some of them occurred in deep water sites. This is probably because there were no Precambrian animals adapted for grazing, one of the standard ecological niches in living marine faunas. Living stromatolites owe their existence to special conditions inhibiting such grazers. It is not altogether fanciful to visualize vast stretches covered with stromatolites, enduring for a period of considerably more than 2000 Ma of the Precambrian.

Cyanobacteria and algae produced free oxygen during the process of photosynthesis. This process was probably the most important ingredient for setting up the conditions for the evolution of all higher organisms. The photosynthetic activities of cyanobacteria and algae proceeded uninterrupted for more than 2000 Ma, and if they were as widespread as seems likely they were capable of adding oxygen to the atmosphere. Little by little an atmosphere was created in which other organisms could breathe; they made the environment fit for animals. It is possible to trace the gradual enrichment of the atmosphere in oxygen by changes in the types of sedimentary rocks that were laid down during the Precambrian. In the earlier part of the Precambrian when oxygen-poor (reducing) conditions prevailed, there are widespread deposits of a distinctive kind of banded iron ore, which could only form when the atmosphere was low in oxygen (as this allows far greater solubility of iron in seawater). These banded iron ores are almost never found in rocks younger than 1000 Ma; the few younger occurrences are small and isolated. There must therefore have been a critical point when there was enough oxygen in the atmosphere to permit respiration so that it was possible for animals to exist, and

above: Late Precambrian Bitter Springs Chert (about 850 Ma old) has yielded a variety of micro-organisms, including the spiral *Heliconema*, with cells apparently in the process of division.

perhaps begin feeding on the plants that had made their existence possible. At the same time the atmosphere would have acquired its protective ozone layer which would cut out some of the more harmful effects of solar radiation.

Cherty rocks from the Precambrian have now yielded many remains of cyanobacteria and other bacteria, many of which have been discovered in the last 20 years. Most of the early species reproduced by fission, that is, their cells simply divided into two, so that the daughter cells were identical to the parent cell. Not surprisingly this mode of reproduction leads to large numbers of cells very quickly, but their capacity for change is limited. Any spontaneous change (mutation) in the genetic make-up of the cell is likely to be disadvantageous, sometimes even fatal. There is evidence from experiments with living bacteria, however, that advantageous changes can and do happen, and that when such a change occurs its spread into a whole population is extremely rapid.

From the late Precambrian Bitter Springs Chert of Australia more complex cellular arrangements have been found. Because these fossils preserve some of the details of the inside of the cells, their preservation must have been extremely rapid. We have to visualize the cells being impregnated with silica as they were still alive, in a similar way to that of the Devonian Rhynie Chert preserving the fine details of much later plants (see Chapter 1). Cherts with ages of about 1000 Ma have yielded the remains of cells which seem to be undergoing cell division of a different kind from simple fission. In these cells there is evidence of the combination of two cells from different plants of the same species, with subsequent division and sorting out of the genetic material from the two parents. Some miraculously preserved specimens even seem to show the nuclei that carry the genetic material in the process of undergoing this kind of division. These fossils are particularly important because they show that plants had evolved different sexes at this time. Cross-breeding had several important results, but in particular it increased the chances of the inheritance of advantageous changes. It would have provided a stimulus for further evolution. Recent scientific discoveries have yielded the first record of sexual differentiation in an alga called *Bangiomorpha* at about 1200 Ma ago.

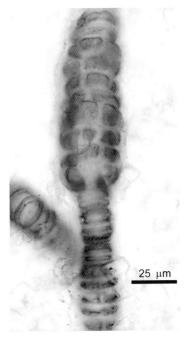

25 μm

There is evidence to suggest that the open oceans were colonized at a similar time. From at least 1000 Ma ago onwards, some of the more oceanic sediments that were deposited in the Precambrian have yielded tiny single-celled algae called acritarchs. If the oceans then had something like the dimensions they do today it would be difficult to overestimate the importance of these tiny plants, for their photosynthetic activity would have greatly boosted the quantity of atmospheric oxygen. They were also the foundation of a food chain. Even today most of the animals that live in the ocean depend ultimately on minute

above: Filaments of *Bangiomorpha pubescens.* The oldest documented case of sexual differentiation in the fossil record is from this species.

photosynthetic algae, which provide food for the minute planktonic animals, which are in turn food for larger fish. Some scientists believe that the Precambrian ancestors of most of the major animal groups were small planktonic animals, but unfortunately there is virtually no fossil evidence of these tiny animals. There is no doubt that they would be extremely difficult animals to find in the fossil state, because of their lack of hard parts and small size, and we can only hope that somewhere they have been preserved by the kind of geological miracle that allowed the preservation of the early plants and bacteria.

In many accounts written not so many years ago (and regrettably in some cases only just published) it is common to find the Precambrian described as 'lacking fossils'. This is obviously untrue. Of course, many Precambrian rocks are devoid of organic remains because they have been metamorphosed. However, in many of the rock sequences that have escaped metamorphism some organic traces have been discovered. So far the emphasis has been on the bacteria and algae which served to prepare the world for animals. The later part of the Precambrian is known as the Proterozoic era, with a status equal to that of the Palaeozoic which follows. No doubt as more people look for fossils in these ancient rocks more will be discovered. The latest part of the Proterozoic, the Vendian (or Ediacarian), has been the subject of particularly intense activity in the last 20 years.

Although the earliest history of the animal phyla is still obscure, there are now a number of good animal fossils from the Vendian, and as the search is intensified more seem to be turning up each year. It is still true to say that almost none of these Precambrian animals had hard parts — shells, bones or the like — and so the circumstances in which they have to be preserved are somewhat special. Conditions in the late Precambrian were different from those that followed in the Palaeozoic, and some of these conditions may have favoured the preservation of soft-bodied animals. For example, it is quite likely that scavenging animals had not evolved at this time, so that the bodies of dead animals were permitted to lie around for longer than they would at the present day, thus increasing their chance of preservation. Bacterial mats on the sediment surface may have facilitated special preservation. Some of the early animal fossils are of various kinds of jellyfish, and their stomach cavities are occasionally preserved as casts in fine sediment. A few of these jellyfish may even be related to forms living today, and it is reasonable to assume that they had similar habits, drifting about in the open oceans carried by currents, feeding on planktonic organisms. Their stratigraphic position below the base of the Cambrian has been known for some time; *Brooksella*, a jellyfish from the Precambrian rocks at the base of the Grand Canyon sequence was discovered many years ago, while some of the Canadian examples have been known for almost a century. It was some time, however, before they were accepted as 'real' fossils. Even now, some of the supposed Precambrian jellyfish have been shown to be a rather peculiar kind of sedimentary structure, entirely inorganic, produced by the expulsion of water from wet sediments. However, more and more examples are now being found, many of then indubitably true fossils, and their worldwide distribution is impressive: several localities in Canada, the United States, Australia, Russia,

above: Fossil jellyfish from the late Precambrian Ediacara fauna, South Australia.

and scattered occurrences in Europe, including Britain. We might anticipate that a free floating organism like a jellyfish would have a wide geographical spread, but it is surprising to find that so many localities had the right sedimentary conditions for their preservation. It is clear that the late Precambrian oceans supported a variety of planktonic single-celled algae, many drifting jellyfish, and a variety of small, soft bodied planktonic animals, perhaps resembling the larvae of living marine organisms, which provided food for the jellyfish.

There are also a few late Precambrian fossil faunas that include a much wider variety of fossils than simple jellyfish. The most famous of these is probably the Ediacara fauna from southern Australia, where there is exceptional preservation of many soft-bodied organisms. The true affinities of these impressions is still under debate. It is possible that segmented worms (annelids) are represented by such fossils as *Dickinsonia*. Some other curious segmented animals have been suggested as arthropods, but

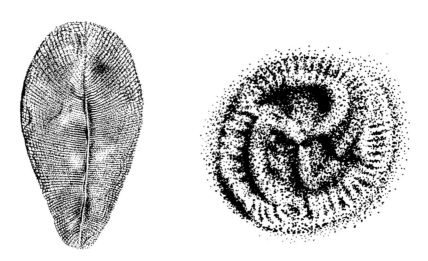

above: Fossils from the late Precambrian Ediacara fauna from South Australia. The possible annelid worm, *Dickinsonia* (left), and the enigmatic *Tribrachidium* (right).

this is far from generally accepted. If it is true that the arthropods were derived from a segmented, worm-like animal, then we might expect to find the common ancestor of all the arthropods in the later part of the Precambrian period. The peculiar animal *Tribrachidium*, with its three curved grooves, has been suggested as an ancestor of the echinoderms, but there is really not much evidence to support this. A common element in the Ediacara fauna, and one which has been found in other parts of the world, including Newfoundland, Russia and Charnwood Forest, England, are frond-like fossils which look rather like the living sea pens. These animals have attracted a lot of controversy about their true affinities. It is no simple matter to relate soft-bodied Precambrian fossils to living animal phyla, and the Ediacara fauna has shown that there seems to have been a distinctive late Precambrian fauna with a number of curious animals all of its own. This impression is confirmed by the discovery of other faunas in Russia and Newfoundland which include some forms in common with the Ediacara, but have others unique to them. The Newfoundland fauna occurs in a series of beds exposed at Mistaken Point, a promontory on the south of the Island. Here, centuries of weathering have etched out the impressions left by the soft-bodied animals, which are left like a picture gallery of the distant past, now washed by the waters of the present Atlantic Ocean. The age of the Ediacara fauna is not much older than the base of the Cambrian, but the Newfoundland faunas may be rather older. The peculiar animals were evidently widespread, although rarely preserved as fossils, and may have held sway over the late Precambrian sea for perhaps 50 Ma.

Apart from the jellyfish, most Precambrian animals were probably bottom dwellers. It has been suggested that they housed symbiotic bacteria in a 'Garden of Ediacara'. Even in rocks where the remains of soft-bodied animals are not preserved you might expect to find the traces left behind by the activity of animals in the sediment — burrows and trails. Oddly enough, trace fossils are relatively rare in Precambrian rocks, even though they are common in rocks of early Cambrian age from many parts of the world. A number of supposed trails have been found in Precambrian rocks (in excess of 1000 Ma old), and some of these may be genuine. However, similar looking marks can also be produced by inorganic causes and so whether you interpret them as animal or not depends on how much you want to believe that there were animals around at the time. In some sections of rock, in Arctic Norway for example, it is noticeable that the numbers and variety of trace fossils seem to increase as the boundary with the Cambrian is approached. Once the boundary is passed it is easy to find deep burrows, vertical to the sediment surface. Perhaps it is true that deep burrowing habits were only acquired by animals of Cambrian and younger age.

CAMBRIAN ORGANISMS

No survey of the early history of life would be complete without taking the story over into the lowest part of the Cambrian period. The old idea that life appeared quite suddenly at this boundary will by now be seen to be entirely wrong; we have traced the gradual increase in variety of organic remains through the Precambrian from about 3500 Ma ago. The evidence is still tantalisingly incomplete and it is still possible that one discovery which miraculously preserves a whole fauna and flora could change the story in many details. The last 30 years have seen a great increase in the effort put into studying fossils from the earliest Cambrian. Although Cambrian rocks were first discovered in Wales, the earliest beds of the system are not well-developed there, and the most important rock sections for studying this interval of time are from other parts of Europe, China, Morocco, Australia, Canada, Greenland, the United States and Siberia. These are mostly sediments deposited in relatively shallow water marine environments. The best approach to understand what happened in the very earliest Cambrian is to collect fossils bed by bed through the critical interval, charting the first appearance of the various kinds of animals the rocks contain.

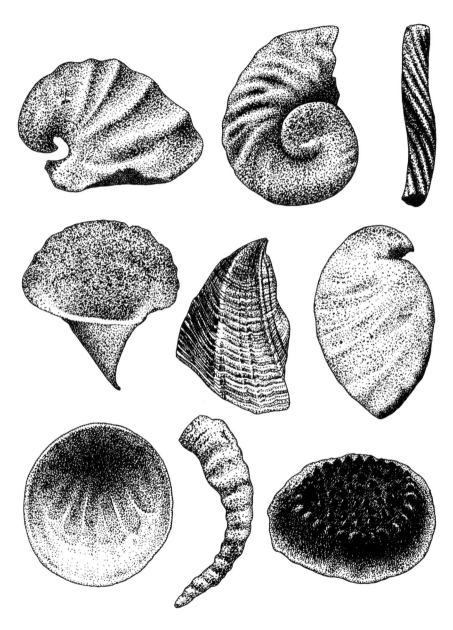

above: Some of the oldest fossils with hard parts that appear near the base of the Cambrian; few of them can be easily related to living organisms (all a few millimetres, less than 0.5 inches across).

It is a remarkable fact that animals with hard parts appeared widely at the base of the Cambrian. For several thousands of millions of years there had been life, but no organisms had secreted a shell, or any kind of preservable strut to serve as a skeleton. Within what must have been a comparatively short time period, many different kinds of organisms became able to do just this. It is still something of an enigma that the appearance of these hard remains from so many different sections in different parts of the world was at approximately the same time. What is certain is that it happened within a very short time period about 543 Ma ago. Not all of the various kinds of animals that are characteristic of the Cambrian made their appearance together in the earliest rocks. There was a time lapse between the very first hard parts to appear in the rocks, and the appearance of some of the animals which are familiar in later Cambrian and Ordovician rocks. Trilobites are not present among the earliest shells, for example. The organisms that appear first are peculiar small shelly fossils, many of which are geological enigmas. They simply cannot be fitted conveniently into the groups of animals we know about from younger rocks. A lot of them are simple tubes, sometimes twisted, and with differing cross-sections. They could have been secreted by any number of different worms, or by something else. There are a number of odd, cap-shaped shells, which are even more puzzling. Some of them appear to fit together in a kind of mosaic, and very few of them appear to be related to anything else that follows. Yet some of them are also very widespread, so whatever kind of animals produced them, they were evidently successful for a short time. In some sections these peculiar assemblages of fossils precede the rocks that contain more familiar kinds of animals, like trilobites. Perhaps these early oddities were 'experiments' that had a short heyday, but did not survive long once the typical Cambrian faunas were established. It is also clear that several kinds of small shelly fossils may have belonged to a single, larger animal. The discovery of an entire specimen of the enigmatic animal

above: One of the earliest trilobites from the Cambrian, already quite a complex animal with well developed eyes.

Halkieria from Greenland has shown that many completely different shells were housed on a much larger animal.

Among the early Cambrian fossils were the peculiar archaeocyathids, a distinctive group of sponge-like forms which produced the earliest structures that might be called reefs. They made their appearance along with the earliest faunas with hard parts, and did not long survive the early Cambrian. The kinds of limestone that contain archaeocyathids also yield remains of calcareous algae that have secreted a calcium carbonate skeleton around their delicate threads, so it was not only animals that acquired the ability to build skeletons at this time. Trilobites regularly appear among the early Cambrian faunas, and often there are a number of different species together. The earliest trilobites were already quite complex animals with

left: *Helicoplacus everndeni* from the early Cambrian, Westgard Pass, California, USA.

well-developed eyes. The earliest molluscs are also present in early Cambrian rocks, and include representatives of the gastropods, chitons and the monoplacophorans, but many of these are hard to trace below the

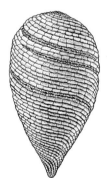

above: Restoration of *Helicoplacus*; a peculiar and puzzling Cambrian echinoderm.

Ordovician. The earliest echinoderms from the early Cambrian include some very odd animals that are not easy to relate to any of the echinoderms that are prominent in recent oceans, most of which have histories going back to the Ordovician. The novelties include the extraordinary helicoplacoids, which, like so many of the more peculiar animals, did not survive past the early Cambrian. Many of the animal groups which dominate Ordovician and younger rocks have only the sketchiest Cambrian histories — admittedly we can recognize other representatives of the same phylum, but these tend to be rather odd animals, with a history confined to the Cambrian.

There are further peculiarities of these early faunas. A number of early Cambrian animals, including many of the odd ones, seem to have used calcium phosphate as the material to build their skeletons. Among the animals that dominate Ordovician seas right through to the present, calcium carbonate is by far the dominant building material for skeletons (with the important exception of the vertebrates). Two kinds of animals used silica for the construction of their skeletons: some of the sponges and radiolarians. These two groups also have a history extending back into the Cambrian. When skeletons were first produced, it seems as if there were no particular advantages in using one material or another, but subsequently calcium carbonate proved more advantageous. Whether there were any conditions that particularly favoured the use of calcium phosphate in early Cambrian times is more difficult to say, although explanations have been put forward which relate phosphate deposition to the gradual changes that were happening in the seas and atmosphere, as conditions became more like those today. In any case, the common occurrence of phosphatic skeletons is yet another reason to regard the early Cambrian faunas as distinctive.

So we come to the critical question: why did the appearance of hard parts apparently happen within such a short time? In the past, before so much was known about Precambrian animals, the answer might have been that there were no animals until shortly before the Cambrian. The sudden appearance of the various animal phyla would then have naturally been accompanied by the sudden appearance of their shells. Today, there are advocates of such a Cambrian 'evolutionary explosion' who maintain that the faunas of Ediacara type fell victim to a major extinction, and that the Cambrian appearances were the result of a subsequent period of frantic and accelerated evolutionary change. The alternative view emphasizes the fact that even in the earliest part of the Cambrian, the types of organisms with shells were already highly varied, showing that evolution had been proceeding on a number of independent lines for a long time. It would be hard to believe that such independent lineages should by chance alone 'decide' to place a premium on acquiring hard parts at the same time.

It may be possible to fit the Cambrian events into the broader picture of the evolving atmosphere, and the changes in the Earth as a whole, which were discussed earlier in this chapter. There is some evidence to indicate that it was not possible for calcium carbonate (or perhaps the other skeletal minerals) to be deposited by living tissues until the partial pressure of oxygen in the atmosphere had reached a critical level. Possibly this level was reached near the base of the Cambrian (although many experts place this event much earlier). If conditions suitable for phosphate deposition had been reached slightly earlier, this might explain why there were so many phosphatic shells at this time. There was also an exceptionally widespread glaciation in the late Precambrian which spread into tropical latitudes. There has been the suggestion that the sudden appearance of ancestors of animals with hard parts may have been related to this event. Glaciation would have lowered sea level, and the cold climate may have produced effects dramatic enough to exterminate the earlier animals lacking shells that dominate the Precambrian. Subsequent elevation of sea level following the melting of the Precambrian ice caps would have produced a worldwide marine flooding, favouring the evolution of new animal designs.

The base of the Cambrian thus marks this change in environmental conditions, and most of the earliest Cambrian animals seem to have been inhabitants of the shallow seas that transgressed across the Precambrian landscape. This recolonization may have come from planktonic animals (or perhaps the deep sea) originating in areas protected from the glacial crisis. Planktonic animals and plants, including jellyfish and single-celled algae, would have been best adapted to surviving the peak of the glaciation by retreating to the open ocean, and it is indeed true that of the Precambrian organisms these two groups alone pass into the Palaeozoic and beyond. If this explanation is correct then the earliest Cambrian would have been a period of extraordinarily rapid evolution — an 'evolutionary burst' like the radiation of mammals in the early Tertiary. Perhaps the ancestors of many of the early Cambrian forms were planktonic animals, and perhaps it is not coincidence that all major phyla still have planktonic larval stages. The growth of the shell may have been a response to settling on the sea bottom. All this is speculative, and very difficult to test one way or the other. The acquisition of shells and skeletons is one of the great milestones in the history of the biosphere, and the difficulty of finding a single neat explanation only adds to its fascination. Recent evidence from molecular sequence modelling of a variety of genes exploring the divergence of phyla and classes of animals, does point to a longer Precambrian history than the fossil record indicates. If this is confirmed it means that there are still Precambrian fossils to be discovered which may change our ideas of the early biological history of the planet.

above: Probable monoplacophoran from the early Cambrian, New Brunswick, USA.

Chapter Six

EVOLUTION
AND EXTINCTION
· ·

EXTINCT animals had their day, were somehow unfit for the modern world, and as a result, died out. This chapter explores some of the causes of extinction, and shows that without extinction there would have been little chance for evolution. While the attitude of conservationists to prevent the wanton destruction of species is wholly laudable, the fact remains that in the normal course of events species will become extinct. The fauna of the present day is the product of repeated changes in the faunas of the past, and there is no reason to suppose that the process will stop now that *Homo sapiens* covers much of the world, although there is equally no reason why humans should speed up the process by wholesale destruction of habitats or excessive hunting. As the world has changed, so have the faunas and floras, with evolution and extinction playing their complementary parts in constantly reshaping the biosphere.

opposite:
Neanderthals were short and stocky, well-adapted to the rigours of life in Ice Age Europe 50,000 years ago. The brains were comparable in volume with modern humans.

There are two types of extinction. In one, the process is as final as in the case of the dodo. A species dies out completely, leaving no progeny. If a whole group of related species become extinct at about the same time, a major animal group may become extinct. There are no dinosaurs lurking in remote corners of the world, not even in Loch

Ness. More than anything else it is the disappearance of such major groups that has changed the appearance of the fauna through geological time. A second type of extinction involves the generation of new species. One species gives rise to another, by any one of a number of processes. Ultimately the parent species may become extinct, but in a sense its genetic material lives on in the daughter species. The only totally dead groups are the side branches of the evolutionary tree, for which the long chain that connects the living species with the ancestors from which they sprang has been irrevocably severed.

EXTINCTIONS AND THE FOSSIL RECORD

If extinct species, which are the ancestors in distant geological periods of living fauna, have hard parts and are readily fossilized, it should in theory be possible to read the past simply by collecting fossils through the strata from the youngest rocks to the oldest. Anybody who tried to unravel the evolutionary story directly from the rocks would be disappointed. As Darwin was well aware, the rocks seem to have many gaps and holes in the record. The ancestor hardly ever seems to be sitting there, where it should, in the rocks immediately below the descendant species. Sometimes a species thought to be close to the ancestor of a whole group of animals turns up in surprisingly young strata. The fossil record is imperfect and capricious. This has led some palaeontologists to reject the stratigraphic order in which fossil species occur as being of no significance at all in the elucidation of the evolutionary process. It is certainly not surprising to find primitive-looking animals in rocks younger than we might expect; after all, the 'living fossils' which are important in unravelling the relationships between different animal and plant groups are all in a sense organisms that have out-lived their time. Some primitive animals are well adapted to their own ecological niches from which they have never been displaced. Evolutionary events do not happen in a regular, step-wise fashion; it is a rather messy, irregular process, with survivors persisting alongside novelties. Momentous evolutionary steps are often taken by humble organisms, while the dominant animals may undergo spectacular evolutionary extravagances that are doomed to total extinction. Some theorists have even said that almost the whole pattern of evolution may be accounted for by chance alone.

Why does the fossil record seem to be so imperfect? Part of the reason is that there are many breaks in even the most continuous-looking rock succession. Bedding planes are the record of such breaks. In some successions deposition occurs irregularly, such as after a major storm, and estimates suggest the rock may represent as little as 1% of the total time in some sedimentary sites. The evolutionary process is relatively slow, and one might expect to find the evolutionary story visible even through small pieces of the record, just as you can recognize the subject of a *pointilliste* painting by standing at a distance. A more important reason for imperfections in the fossil record is that most rock successions record shifts in facies: the sedimentary (and biological) environment changes with time, and as this happens the evolving animals are carried elsewhere and the descendant species are recovered from a different rock succession, maybe many kilometres away.

Perhaps the most significant factor affecting the fossil record stems from the way new species are derived. Where the generation of new species has been studied among living animals a most important cause of a new form is the isolation of a population at the fringe of its range. These fringe populations become separated, change in response to some slightly different set of conditions and eventually become different enough from the parent stock so that they cannot interbreed, i.e. a truly independent biological species. A shift in conditions may allow the descendant species to spread out and even displace its progenitor. Because the inception of a new species happens at the edge of a population, the chances of finding the actual site where it happens preserved in the rocks is rather low; the point of origin is nearly always somewhere else. The result in any particular rock section is a rather patchy mosaic of species that all seem to be related, and in some cases may record actual ancestors and descendants, but which can certainly not be read 'like the pages of a book'.

Added to this is another kind of change which can be recorded in sedimentary successions. This is a slow drift of change which occurs within a species through time, a shift in shape perhaps, or an increase in size, which cannot be explained by isolation of populations. In this case the ancestral species is transformed into the descendant one and so does not

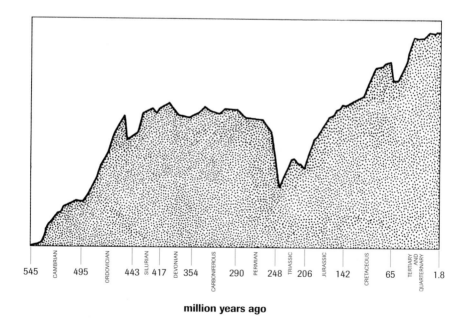

million years ago

above: Numbers of major groups of marine organisms through geological time. After an early climb through the Cambrian, the numbers are rather uniform except at the periods of major extinction, especially at the Permian–Triassic boundary.

really become extinct in the normal sense of the word. Some palaeontologists hotly dispute that this kind of change represents evolution at all, and there is a lot of hair-splitting about whether or not the descendant is really a different species. Regardless of the theoretical arguments, there are many rock sections in different parts of the world where this kind of change has been seen. It is of practical importance, too, because these kinds of small changes often occur in the commonest fossils, and become the basis of stratigraphic zones for the precise dating of rocks by their fossil content.

This complicated introduction is necessary to understand the complexity of the fossil record. For example, if the new species arose by the kind of isolation of marginal populations that is important today, then we might expect the fossil record in any limited area to have gaps and jumps in it. In fact, the fossil record might not be so incomplete as has been thought; the jumps may be a natural phenomenon.

The generation of new species and the extinction of others has been a continuous process since the Cambrian. If one considers life in the seas, there was a steady growth in the total number of species during the Cambrian, and the number has remained high since then. But there have been a number of important times when extinction greatly exceeded the generation of new forms, and these were the times when whole groups, such as the rugose corals or the ammonites or the dinosaurs passed from the world forever. Very important mass extinctions occurred at the end of the Permian and the end of the Cretaceous periods, and it is no coincidence that they also define the end of the Palaeozoic and Mesozoic respectively. These were great biological crises that changed the faunas of the Earth. Here we have to invoke more powerful causes than the usual processes of change described earlier in this chapter, and these will be discussed later. But first it is necessary to describe

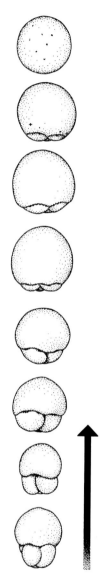

above: Gradual transformation of the foraminifera *Globigerinoides* into *Orbulina* (Miocene).

the kind of small scale evolutionary changes that are the bread-and-butter of the fossil record.

THE MOSAIC OF EVOLUTION

The evolutionary story is best preserved in the least conspicuous of fossils. More momentous changes have often taken place in sites where the fossils stand a low chance of being preserved, however, there are now fossils that serve to connect birds with dinosaurs, even if some intermediate steps are still lacking. To see the record at its fullest one has to turn to sites where there is every reason to suppose that sedimentation has been continuous. One such place is in the deep sea, where the steady rain of plankton carries on regardless of the shifts in sea level that affect the continents, and the sea bottom is a graveyard for the rain of tiny organisms that die nearer the surface. Some of the small planktonic foraminifera record the kind of continuous changes that often evade the palaeontologist elsewhere. Minute species of the genus *Globigerinoides* change step by step into a perfectly spherical form known as *Orbulina*. The appearance of this spherical animal serves to define the base of the mid-Miocene period. The same sequence of changes from the one species to the other has been found repeatedly in many sites all over the world. The cause of the change is a mystery, but whatever the reason the effect is of the greatest use in dating rocks of Tertiary age.

This kind of simple change can be matched by other examples from planktonic foraminifera, and it seems to be characteristic of planktonic organisms in general. Graptolites also show continuous changes of this kind. In the earlier half of the Ordovician, slender 'tuning fork' graptolites of the genus *Didymograptus* show a progressive change to stouter, longer forms, while the v-shaped species of *Isograptus* produce longer, more robust kinds of colonies. Ammonites also tend towards a continuous sequence of changes, sometimes with the multiplication of ribbing or with progressive changes in the outline of the shell. In these planktonic or pelagic organisms it seems that whatever advantages were conferred on the species by the change were transmitted to the population as a whole.

Among bottom-living fossils it is perhaps more usual to find rather sudden jumps between one species and the next, probably because the species evolved by geographic isolation in the manner described previously. The sea urchin *Micraster* is one of the commonest fossils in the European Chalk (Cretaceous), where its heart-shaped tests have acquired the common name of 'shepherd's crowns'. The Chalk seems to have accumulated as a pure, lime ooze, largely formed by the microscopic remains of minute algae, and foraminifera. Although not a sediment of the open ocean, its record of deposition is remarkably continuous. Changes in the sea urchins record an evolutionary story connected with burrowing habits. As the sea urchins acquired the habit of burrowing deeper in the soft, chalky sediment their tests underwent a series of small changes: the entire animal became higher and inflated, the plates that make up the petal-like ambulacral area became

above: Changes which occurred in the heart urchin, *Micraster*, during the Cretaceous.

more inflated, and the lip around the mouth became exaggerated. The series of species leading from *Micraster corbovis* through *M. cortestudinarium* to *M. coranguinum* record these changes, and the same species may be collected in the same order across much of Europe. They have become important in determining the zones of the Chalk. In this case it is possible to correlate the changes in the fossils with their mode of life, by comparison with living sea urchins which show similar changes in relation to life habits. As the animals learned to burrow more deeply it became necessary to retain contact with the surface for breathing through a long tube, kept clear of sediment by extended tube feet. The points of attachment of these specialized tube feet became modified in a special way as the tube become longer. The *Micraster* species were not the only organisms to show slow, ordered, evolutionary change as the Chalk was slowly deposited; large molluscs of the genus *Inoceramus*, as well as tiny foraminifera that make up much of the sediment also changed, and all of their changes act as a chronometer for the passage of Cretaceous time.

The Chalk is an exceptional deposit in its continuity and uniformity, without the kind of drastic change in sedimentary facies that introduce problems into more usual sedimentary sequences. However, even the Chalk is not without its changes; there are beds where deposition temporarily slowed down or ceased, called 'hardgrounds' because the sediment surface became crusty. Here a whole series of animals that are normally rare in the Chalk become numerous, and unlike the sea urchins their evolutionary history cannot be deduced from careful collecting in the beds above and below, simply because they cannot easily be found there. With many fossil groups this is the normal state of affairs; fossil remains are found in certain special horizons, some of which become justly famous for the beauty of their preservation. They tell us much about the animal species concerned, but nothing about the history of the animal in the time immediately before or after. Usually the gaps betweenthefossil-rich horizons may be measured.

in millions, if not tens of millions of years, and they are often widely separated geographically. It would be a mistake to think we could stack these fossils up in their time relationships to give us a picture of how they changed, in the manner of the sea urchins. There are too many ambiguities in the gaps. The best that can be done in these circumstances is to look very carefully at the features they show, decide which ones were derived during the course of evolution, and construct a diagram to show the progressive stages in their pedigree (technically, a cladogram). For most fossils the construction of cladograms is the best we can do. As usual, it is a matter of debate how imperfect the record of a particular group of organisms is, and different scientists have differing views about how important the stratigraphic order should be.

HUMANS

'The proper study of mankind is Man.' The resources put into the hunt for the fossil remains of humans are proof of Alexander Pope's aphorism. The search for palaeontological evidence for human derivation has occupied more newspaper space — and provided more scandal — than all other fossil discoveries. Since Darwin published *The Descent of Man* in 1871 the idea that humans and the higher apes were closely allied has been lodged in the common consciousness. Definitive fossil proof was lacking until recent decades. It was hoped that there would be a single missing link lodged in the rocks — a perfect amalgamation of human and ape characteristics that would prove our ancestry at a stroke. Such a fossil resolutely refused to turn up. So eager were the scientific community for such a discovery that when the Piltdown Man skull was unearthed in 1912, it was quickly accepted by many respected workers as the answer to the 'missing link'. Here was a skull that seemed to display an amalgam of ape and human characteristics. Sadly, that is exactly what it was, an elaborate forgery composed of bits of human and bits of ape, carefully matched to look genuine. It was exposed as a fake when the different ages of the different pieces were proved, and in particular the recent ages of the human and ape 'fossils' were demonstrated by radiocarbon dating. Even now controversy still rages over who was 'in the know' about the forgery.

Gradually, other genuine pieces of human-like apes (or ape-like humans) began to turn up from areas well-removed from Europe, notably China, Java and Africa. It has been the African continent more than any other which has yielded the fossils that have provided not one missing link but a whole chain of intermediate forms and 'first cousins', showing that the evolution of humans was much more complex than had been supposed. New discoveries of fossils of hominids (Hominidae is the family that includes humans and their ancestors, and according to some workers, chimpanzees and gorillas as well) are being made every year, and the kind of sites occupied by our distant ancestors have been identified in the deposits formed in and around ancient lakes, rivers and caves. Many of the exciting discoveries of the 1950s and 1960s were made by Louis and Mary Leakey, and the family tradition has been continued by their son Richard. Opinions about the meaning of the fossils seem to change with every subsequent discovery, but gradually some sort of consensus is emerging, especially since genetic information has provided independent evidence about ancestry.

Important fossil finds have now been made in many localities, notably in Olduvai Gorge and Laetoli in Tanzania, and in South Africa, Kenya, and Ethiopia. Ancient fossils, probably lying close to the common ancestor of humans and chimpanzees, are known as *Orrorin tugensis* (approximately 6 Ma old) and *Ardipithecus ramidus* (approximately 4.5 Ma old). Some ape-like hominids are placed in the genus *Australopithecus*, and others, which were initially supposed to have been our ancestors, have now been recognized as a side branch: the genus *Paranthropus*. The skulls of *Paranthropus* are unusually flat faced, and the brain case is obviously much smaller than in any human. They lacked the prominent canine teeth that make the gorilla such a fearsome looking animal, and in this respect are more like our own ancestors. They have been found in deposits as old as 2.5 Ma and up to about 1 Ma old. The genus *Australopithecus* includes the species *Australopithecus afarensis* (a remarkably complete skeleton of which was christened 'Lucy') which is believed to represent a species close to our ancestry.

Younger *Australopithecus* and *Paranthropus* species may have been contemporaries of our early ancestors. *Paranthropus robustus* appears to have been adapted to a dry, savanna type of climate. A more slender species of *Australopithecus, A. africanus*, which is found in deposits even older than those with *Paran thropus robustus* (but overlaps with these in time) is often considered to be closer to the main line of human evolution, but its place as the ultimate ancestor of our living species is still open to debate. Whatever the position of *A. africanus*, it is clear that by about 2 Ma ago the first species that can be placed in the human genus, *Homo*, had not only evolved, but were manufacturing tools for complex tasks and probably living in small communities, so were displaying the kind of co-operative habits we like to associate with ourselves. These species are known as *Homo habilis* and *Homo rudolfensis*. They do not have the kind of cranial capacity typical of modern humans, but their brains were larger than those of the

above: Skeleton of *Australopithecus afarensis*, a primitive African hominid that lived 3 to 5 Ma ago, nick-named 'Lucy'.

above: Reconstruction skulls of early and modern humans. In chronological order (left to right): *Australopithecus; Homo habilis; Homo erectus*; Broken Hill man; Neanderthal man; Modern man.

australopithecines (this does not necessarily mean that they were more intelligent). Some remains of limbs that have been found are more like those of humans than those of either apes or *Australopithecus* and *Paranthropus*, but there has inevitably been the usual debate about whether these are our ancestors, or first cousins on a side branch. *Homo habilis* and *H. rudolfensis* do appear more closely related to us than are the australopithecines. Up to this stage the history of the family of humans has been recovered from fossils confined entirely to the African continent.

Homo erectus, by contrast, was much more widespread. Remains of this species have been found in China, Java and western Asia as well as in Olduvai and Eritrea in Africa. In age they more or less follow those of *H. habilis*, falling in the range from about 1.8 million to less than 700,000 years old. The most likely explanation of the widespread distribution of *H. erectus* is that the species spread from the African continent over a large part of the Old World, moving up into temperate latitudes. This must have provoked new responses from the dispersing populations. It is likely that some *H. erectus* populations regularly used fires, and others developed more sophisticated stone tools, including hand axes. Some fossils of *H. erectus* from Africa may be older than those from the rest of the world, which is what one might expect if the invasion idea were correct, but fossils from Georgia and Java may be almost as old. In shape, *H. erectus* is midway between *H. habilis* and modern humans for many of its features. The brain case is lower than ours, and there is a distinct ridge over the brow and one across the back of the head. Some palaeontologists put more stress on the differences from modern humans, and there is considerable debate about how *H. erectus* fits into the theories, and even how many species are involved. It is clear, however, that skeletal features of *H. erectus* and the remains of subsequent species of *Homo* show a reasonable, if imperfect gradation in several characteristics such as cranial size.

At about the same time as the crucial changes between the species of *Homo* were taking place, the climate of the Northern Hemisphere was passing into the coldest phases of the Pleistocene Ice Age. In 1856 a skeleton was discovered in the Feldhofer cave in the Neander Valley, Germany, which has since given its name to a suite of early human fossils known as 'Neanderthals'. Many of these fossils are between 70,000 and 40,000 years old. The typical Neanderthal has a skull that bulges out at the sides, and a lower cranial shape than modern humans, but the capacity of the brain case matches (or even exceeds) that of living populations. The bones of the rest of the body are generally stouter than ours, and these people had very large noses judging from the configuration of the nasal openings. Skeletal remains of this kind are widely distributed across Europe and extend into the Middle East. These individuals have been interpreted as being specialized to cope with the cold conditions prevalent at the time when the remains of Neanderthals evolved through the later Pleistocene. Even today the limbs of Inuit are relatively shorter than those of Masai tribesmen from the tropics, and the Neanderthals were similarly proportioned to Inuit, although considerably more muscular. Today, scientists generally regard the Neanderthals as a separate species, *Homo neanderthalensis*, rather than as a race as they did in the past. Near the end of the last glacial period they appear to have been displaced by modern humans. There is even disputed evidence of a 'hybrid' between Neanderthal and *Homo sapiens*. Skulls

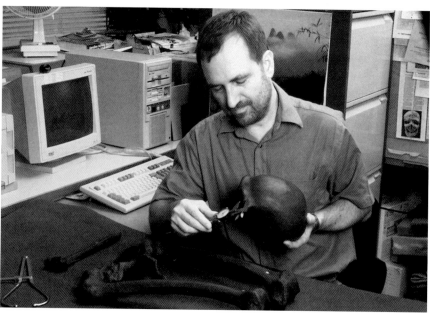

opposite top: This skull of *H. neanderthalnsis* was recovered from Gibraltar, but no associated remains have been found. It was found in 1848, eight years before the more famous Neanderthal find, but lay unrecognised for many years.
opposite bottom: Anthropologist Chris Stringer measures a replica of the skeleton found in the Neander Valley, near Düsseldorf, Germany, in 1856. It gave its name (Neanderthal) to a whole ancient population in Europe.

almost indistinguishable from those of typical living *H. sapiens* are present at about 35,000 years ago (Cro-Magnon), with less complete evidence extending back to 250,000 years ago in Africa. Some of these people had the behavioural characteristics of modern humans, including the propensity for burying their dead — which is why fossil remains become more common in the mid to late Palaeolithic. The use of tools goes back to *H. habilis* and *H. rudolfensis* and probably beyond, but their use and design became progressively more sophisticated after 50,000 years ago.

Probably the greatest change in the understanding of human evolution in the last few years has been the widespread acceptance of the 'Out of Africa' theory. The claim is that modern humans are a comparatively recent innovation, originating in Africa approximately 150,000 years ago, but only dispersing from there during the last 100,000 years or even less. We are all descended from this small population which then migrated outwards (as *H. erectus* had long before), to displace all other humans around the world, and give rise to the modern race. Earlier humans may be recognised as additional species such as *Homo heidelbergensis* of Europe and Africa (500,000 years old), believed to be the common ancestor of *H. sapiens* and *H. neanderthalensis*. Modern humans are younger than was once thought. Genetic studies have given this theory much support as studies of mitochondrial DNA have suggested an African origin. The migration is linked with inferred rapid cultural change at the same time.

In morphological terms it becomes a little difficult to define modern humans. Our complicated patterns of speech are characteristic of humans,

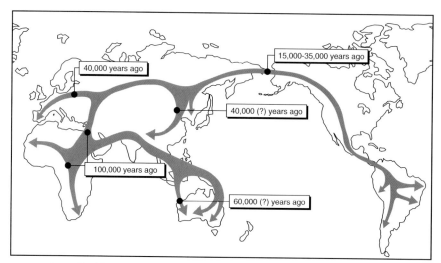

above: The 'Out of Africa' theory, showing how modern man is supposed to have migrated after his African origins.

but the vocal modifications necessary for speech are not uncontroversially recognized from bones. The first complex ceremonial centres and cities ('civilizations') were established by about 9000 years ago, which is, geologically speaking, within the twinkling of an eye from modern humans' appearance about 100,000 years ago. Our species extended still further the global dispersal of the genus *Homo* which began with *H. erectus*. Fossil evidence shows that humans had moved into the New World, via the Bering Straits from Asia by about 15,000 years ago, and possibly into Australia by 60,000 years ago. It is probably this wide geographic dispersal, and the very different conditions this particular species meets over its range, which are responsible for the wide physical variation between races. The adaptability on a cultural scale is unique to humans, and perhaps the best species diagnosis, but it is a product of the mind, and not of fossilizable parts.

Sadly, *Australopithecus* and related species did not survive the Pleistocene. It would be exciting to think of a species of the genus lurking in some remote corner of the world — perhaps as the progenitors of the legend of the Yeti — but it is difficult to believe that any large land animal could have escaped the scrutiny of the 20th century, unless it was equipped with the intelligence of *Homo sapiens*. Whether our own ancestors were involved in the demise of our nearest relatives *H. neanderthalensis* is a matter for speculation, but it would perhaps not be surprising.

TIMES OF MAJOR EXTINCTION

Whether hominid or sea urchin, the kind of evolution and extinction we have described so far has been on a relatively small scale. Splits (or dichotomies) have separated different lineages (for example, *Homo* and *Australopithecus*) one of which may become extinct, but there is no question of the whole of the primates, (the group which includes the monkeys, apes and lemurs as

above:
Neanderthal artefacts from Gorham's Cave, Gibraltar.

well as humans) being threatened with extermination. Even the traumas of an Ice Age, which may produce races or species able to cope with cold conditions (of which woolly mammoths and Neanderthals were examples), do not necessarily stimulate the mass extinction of whole groups of animals. Yet the extinctions that occurred at the end of the Permian and again at the end of the Cretaceous periods were of this kind. These were times of crisis for the whole fauna (and flora) through which only a few groups passed unscathed; many major groups of animals disappeared from the Earth forever. In some cases the extinction was preceded by a general decline in number of families, so that the trilobites had already been reduced to a few forms by the Permian, and their removal before the Mesozoic was only the *coup de grâce*. In other cases the termination was apparently more abrupt — the classic example being the extinction of the dinosaurs at the end of the Cretaceous.

These events seem enormously destructive and yet they allow the subsequent proliferation of other groups: extinction and evolution have been partners in shaping the modern biological world. While the dinosaurs occupied the dominant places in the terrestrial ecosystem there was no obvious opportunity for mammals to become the prevalent large herbivores, but once the dinosaurs were removed the rich herbage of the Tertiary was there for mammals to exploit. In the early Tertiary there was a short time lag after the preceding extinction event to allow recolonization of the available ecological niches, but it was astonishing how quickly the faunas recovered to something like their previous diversity. The two major extinction events are the most pronounced of a large number of phases when extinctions were apparently higher than normal. At the end of the Triassic, for example, almost all the ammonites that had flourished in that period became extinct, a mere two groups surviving to give rise to the different kinds that populated the Jurassic seas. The end of the Ordovician saw the extinction of many of the characteristic trilobite families that make Ordovician fossil assemblages easy to recognize, and the same event affected brachiopods, nautiloids and graptolites. Mass extinctions are also thought to have occurred in the late Cambrian and in association with 'snowball Earth' in the late Precambrian. These extinction events define the ends of many of the geological periods, indeed it was probably the different gross characters of the faunas on either side of the boundaries that enabled the early geologists to recognize the different major periods to start with. The events at the end of the Palaeozoic

plate 34 Fossil horseshoe crab, *Mesolimulus ornatus*, Jurassic, Solnhofen, Germany. Powerful pincers are held close to the body, also equipped with a good 'cutting edge'. Carapace measures 6 cm (2.4 in) across.

plate 35 Fossil bee, *Anthophorites titania*, Miocene, Switzerland.

plate 36 (top left) Crab, *Chaceon peruvianus*, Miocene, Argentina. Width 20cm (8 in).

plate 37 (bottom left) Fossil lobster, *Thalassina anomala*, Pleistocene, eastern Australia. The original shiny cuticle of the lobster is preserved in a yellow, limy matrix. All the legs are attached, some partly concealed. Length about 12 cm (4.8 in).

plate 38 (right) Fossil dragonfly, *Libellulium longialata*, Jurassic, Solnhofen, Germany. Dragonflies have stiff, long wings, with finely netted veining, long slender bodies, and large eyes, all well seen here. Longest wing is 7 cm (2.8 in).

plate 39 Aphid in Baltic amber, *Mengeaphis glandulosa*, Eocene. Length 1.7 mm (0.07 in).

plate 40 Fossil millipede, Carboniferous, Yorkshire, UK. Part and counterpart are shown. Each pair of legs shows as white, hair-like lines. Similar specimens occur at Mazon Creek, Illinois, USA. Length 6 cm (2.4 in).

plate 41 Prawn fossil *Aeger* sp. Preserved in the Jurassic Solnhofen Limestone, Germany.

plate 42 (left) Burgess Shale arthropod, *Marella splendens*. Cambrian, Canada. Length 1.5 cm (0.6 in)

plate 43 (right) Peculiar, screw-like bryozoan, *Archimedes sublaxus*, Carboniferous. While most bryozoans require microscopic examination, a few form colonies large and distinctive enough to be easily recognizable like this with its colonies forming peculiar miniature helter-skelters.

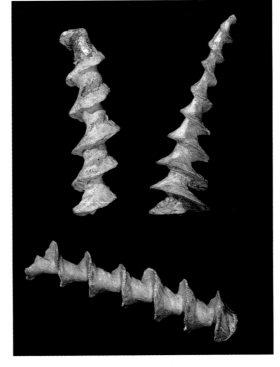

plate 44 Fossil sharks teeth, *Hypotodus robusta*, Early Eocene, Kent, UK. Anterior teeth (tall, slender) and lateral teeth (triangular) are shown.

plate 45 Eocene bony fish, *Pristigenys substriatus*, Monte Bolca, Verona, Italy. It has a deep body, narrowly compressed from side to side, very large eyes, prominent fins and a fan-like tail. Length 10 cm (4.0 in).

plate 46 (top) Extinct ray-finned fish.
plate 47 (middle) *Adriosaurus suessi*, fossil lizard from the Cretaceous of the Isle of Lessina, Dalmatia.
plate 48 (above) *Ichthyosaurus acutirestris*, a marine reptile from the Lower Jurassic of Germany, about 185 Ma old.

plate 49 Lemur, *Megaladapis edwardsi*, Late Pleistocene, found near Ampoza, south west Madagascar.

plate 50 The light, slender dinosaur skeleton of the fast runner, *Hypsilophodon, foxii* from the Cretaceous of the Ilse of Wight, England.

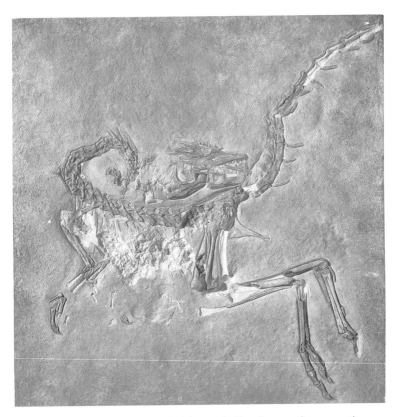

plate 51 A well-preserved fossil of the saurischian dinosaur *Compsognathus* from the Upper Jurassic of southern Germany.

plate 52 *Edmontosaurus regalis* skeleton still half buried in sandstone rock, from the Upper Cretaceous of Alberta, Canada, about 70 Ma old.

plate 53 Fossil frog, *Rana* sp., Miocene, Spain. The skeletal anatomy is preserved in its entirety. The outline of the soft parts are also clear. Specimen 12 cm (4.8 in) long.

plate 54 Skull of the extinct mastodon, *Zygolophodon atticus*, from the Miocene, Pikermi, Greece. This fossil gave rise to the myth of the one-eyed giant Cyclops.

plate 55 Skull of *Diprotodon* skull, the largest known marsupial mammal. This herbivore was over 3 m (9.8 ft) long and lived up to 30,000 years ago in Australia.

plate 56 Fossil root, *Stigmaria ficoides*. This root is preserved as a sandstone cast. Stigmaria is the name given to roots of the giant Carboniferous lycopod Lepidodendron.

plate 57 Fossil lycopod bark, *Sigillaria laevigata*. This Carboniferous fossil is the bark of the giant lycopod Lepidodendron. The specimen is 9 cm (3.6 in) long.

plate 58 (left) Fossil plant, *Archaeopteris hibernica*. The fern-like foliage of an extinct progymnosperm is beautifully preserved in yellow sandstone from the Devonian, Kilkenny, Ireland. Specimen is 25 cm (10 in) long.

plate 59 (right) Fossil wood, probably *Araucarioxylon arizonicum*. Silicified coniferous wood from the Triassic Period. Petrified Forest National Park, Arizona, USA.

plate 60 (left) Miocene maple leaf, *Acer trilobatum*, preserved in a marly limestone. The leaf is divided into a broad central lobe (finely toothed) and two flanking lobes. The leaf is 10 cm (4.0 in) long.

plate 61 (right) Fossil flower, *Porana oeningensis*, Miocene. These beautifully preserved flowers are from the Oeningen deposits, Germany. Fossil flowers are extremely rare because they are shortlived. Diameter is about 2 cm (0.8 in).

plate 62 (left) Fossil fern, *Spenopteris laurenti*, Carboniferous, Derbyshire, England. The frond is preserved as a compression on the surface of a siltstone. Some of the fine detail on the leaflets is preserved. Length 9 cm (3.6 in).

plate 63 (right) Oligocene sumac tree, *Rhus stellariaefolia*, from Eocene rocks in Colorado, Utah and California.

Fossil plant 64 Fossil poplar leaf, *Populus latior*, Miocene, preserved in limestone laid down under freshwater. This species has a finely toothed margin; a large, wide leaf born on a long stem. Length is 18 cm (7.2 in).

and Mesozoic were probably the most dramatic. It has been suggested that the extinction of animal groups at these times may have been because of lower rates of evolution rather than particularly high rates of extinction, such that the normal rate of replacement of extinct animals by new species just did not take place.

LATE PERMIAN EXTINCTION EVENT

At the end of the Palaeozoic the whole group of calcite rugose corals became extinct, to be replaced by the aragonitic Scleractinia in the Mesozoic. Similar drastic changes occurred in the Bryozoa. Most of the typical Palaeozoic brachiopods disappeared, or were drastically reduced in numbers, and only two major groups, the rhynchonellids and terebratulids, prospered in the Mesozoic. Among the crinoids, which were extremely varied in the Permian, only one genus survived into the Triassic, and the blastoids disappeared before the end of the Permian. Nautiloids declined, and the end of the period was a testing time for ammonites and other molluscan groups. Trilobites and other Palaeozoic marine arthropods became extinct, and the Mesozoic marks the inception of most of the important arthropod groups, like the crabs and lobsters that dominate the oceans today. Among terrestrial animals there were also profound effects. Most of the groups of reptiles that dominated the Permian did not survive into the Triassic, though some of these were ancestral to later forms. The great radiation of the archosaurs, including the dinosaurs and crocodiles, dates back to the Triassic and latest Permian. Plesiosaurs and ichthyosaurs have a similar origin in the Mesozoic. Even many important groups of plants, like the cycads, have a history that dates back into the Triassic, and the end of the Palaeozoic marked the decline, if not the end, of the great lycopod trees that had been such a feature of the Carboniferous coal swamps. In short, almost every group of animal or plant was affected by the late Permian event, and the resulting regeneration after the extinction laid the foundations for the modern fauna.

left: The blastoid *Deltoblastus jonkeri* from the Permian of Timor.
right: Star-shaped blastoid *Timoroblastus coronatus* from the Permian of Timor.

LATE CRETACEOUS EXTINCTION EVENT

At the end of the Mesozoic, in the late Cretaceous, the dinosaurs died out. Their dominance over terrestrial habitats that had lasted for more than 100 Ma came to an abrupt end. In a habitat that could scarcely be more different, the open sea, the ammonites also died out, leaving no descendants. The flying lizards (pterosaurs) also took to the air for the last time in the late Cretaceous, and the marine plesiosaurs came to the same end as the ammonites. Even the tiny planktonic foraminifera underwent a revolution at apparently the same time. An event of such magnitude, affecting a diverse selection of organisms in many different habitats, demands an explanation, and there is no shortage of different theories. In the welter of such hypotheses it is important to hang on to the facts, and not get carried away by the most comprehensive or dramatic explanations. It is equally important to remember that certain groups of animals certainly did not become extinct at the same boundary, so at the end of the Cretaceous the lizards and snakes (both small) and the crocodiles (distinctly large) came through apparently unscathed, as did the mammals, which were to inherit the roles vacated by the dinosaurs. We cannot simply bombard the Earth with some sort of fatal dose of radiation, unless we grant some rather exceptional properties to some unexceptional organisms. Also, since the last dinosaurs and last ammonites occur in different kinds of rocks, the one terrestrial, the other marine, there is always the problem of correlating the different events, although now it is virtually certain that the extinctions occurred at the same time.

Not surprisingly, the death of the dinosaurs has generated numerous explanations, but it is a pity that most of these theories have paid little attention to the drastic changes that were going on among all the other groups of animals and plants. One theory suggests that the climate may have become too dry to support enough vegetation to keep the giant herbivores alive; after all, if the herbivores had died out, the carnivores that preyed on them would have become extinct automatically. However, there

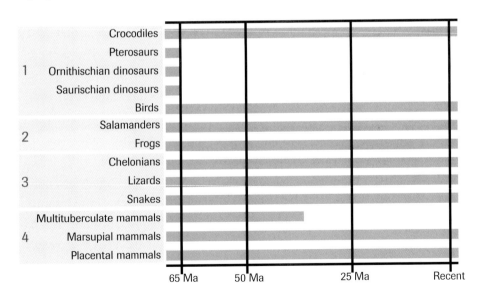

above: The only groups of terrestrial vertebrates that did not survive beyond the end of the Cretaceous period were three groups of archosaurs (1). Amphibians (2), reptiles (3) and two out of the three mammal groups (4) are still living now.

is no real evidence of the sort of widespread desert conditions necessary to induce such a change, which should be reflected in the rocks deposited at the Cretaceous—Tertiary boundary. This explanation does not account for the extinction of other groups (particularly in the sea) or the survival of the crocodiles. If not too dry, then perhaps the climate became too cold, so that the dinosaurs starved to death during the long inactivity of the interminable winter. This period was not one of glacial activity, although there was a climatic cooling at the time, but the effects would have been cushioned in the sea, where events almost as dramatic were taking place.

What about a change of diet? The Tertiary rocks are dominated by flowering plants, and so is it possible that the dinosaurs could not cope with this new kind of sustenance? In this case the timing is wrong; the big evolutionary burst of flowering plants occurred well before the extinction of the dinosaurs. Could the small mammals have caused the extinction of the dinosaurs by developing an inordinate taste for dinosaur eggs (which some scientists believe became thinner shelled in the late Cretaceous)? Possibly, but the trouble with this kind of theory is that there is no way of proving it one way or the other. What makes it unlikely is that there were so many other changes taking place at the same time which had nothing to do with the eating habits of mammals. The mammals could have learned such eating habits in the mid-Jurassic or the early Cretaceous, and it seems unlikely that coincidental changes should have occurred at the same time in different environments. So it goes on: theories are proposed and defended vigorously by their adherents, only to flounder against some inconvenient fact, or belong to the category of 'pseudo-science' for which an explanation, no matter how dramatic, is not capable of being tested against the facts.

Let us return to the facts that are available for the late Cretaceous, either gleaned from the rocks or from our new knowledge about the geography of the time, to see if these can throw any light on the problem. What happens when rock sections spanning the Cretaceous—Tertiary boundary are examined in detail? Are there any clues which might elucidate the cause of the mass extinction? The latest and most persuasive of the catastrophe theories of dinosaur extinction maintains that the extinctions were the result of the collision of an asteroid (or several large meteorites) with the Earth at the end of the Cretaceous. The original evidence, which has been confirmed in many rock sections subsequently, related to an unusually high content of the element iridium, which had been detected in a 'boundary clay' forming the junction between the end of the Cretaceous and the beginning of the Tertiary. This high iridium anomaly was thought to be the product of a major extraterrestrial impact. Geochemists have now investigated many rock sections spanning the critical interval from widely separated parts of the world, and in virtually all of them the same iridium anomaly has been discovered in the same horizon.

Furthermore, other signs of meteoritic impact, such as 'shocked' quartz, have turned up in these sections at the same level. Iridium anomalies can result from causes other than meteoritic impact, but it would take a very sceptical mind to discount this mass of evidence. It does appear that there was a major impact of an extraterrestrial body close to the time when the dinosaurs and ammonites died out, and many other animal groups suffered trauma. In some sections soot horizons have been detected at the appropriate level — just what you would expect if there were a major conflagration. Following on from the iridium and soot layers there is often an increase in fern pollen. This also supports the theory, as ferns are the first plants to recolonize after major environmental disasters.

There are a number of scenarios which describe how the impact of an extraterrestrial body is supposed to have caused mass extinction. Most of them include variants on the 'nuclear winter' — an explanation which was developed in anticipation of the catastrophic effects of a nuclear war. Vast quantities of dust were thrown up into the atmosphere, smoke poured from massive forest fires, blotting out the Sun, and this killed the vegetation which needs sunlight to photosynthesize. There was a similar catastrophic effect on plankton in the sea, which forms the base of all marine food chains. Because the largest dinosaurs were vegetarians they could cope neither with the loss of their food, nor with the cold weather. The carnivorous dinosaurs that preyed upon the grazers soon died out when their prey was no longer available. Survival was a matter of being able to tolerate low temperatures, or being able to enter a state of dormancy (as a seed, or a burrowing larva, for example), or perhaps just having the ability to eat a variety of foodstuffs. The small, warm blooded mammals might have survived on a diet of scavenging insects. In the sea, crabs could have picked from the varied larder they still enjoy today.

This theory has a simplicity which makes it very attractive, if one can apply such a word to one of the greatest disasters in the history of life. It seems almost parsimonious to complain that there are some difficulties with it. The extraterrestrial impact seems very likely — but did it really cause the extinctions? In the first place it is claimed that the animal groups that became extinct, ammonites and dinosaurs, began to decline before the very end of the Cretaceous. If this were so it could scarcely have been in anticipation of an extraterrestrial impact. Second, it is hard to explain why some of the animals and plants that survived were apparently affected so little, for example insects and flowering plants. Cretaceous flowering plants are closely related to living ones, and most of the insects can be placed in living families, or even genera. Surely Cretaceous events would have affected the flowering plants as much as any group of organisms. One moth from the Cretaceous has living relatives that eat pollen, and there is no reason to suppose that its Cretaceous relative did otherwise. How could this moth have survived the darkened years? Similarly, colonial corals in

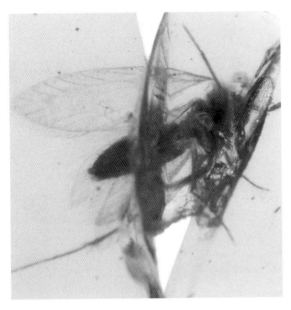

left: *Sabatinca perveta*, a moth belonging to the living pollen-feeding family Micropterygidae, in Burmese amber of late Cretaceous age.

the sea are exceedingly sensitive to light and pollution, yet they were not extinguished either. Already it seems necessary to introduce the notion of an area of the world that was not affected by the catastrophe to the same degree: possible, of course, but more complicated. There are also claims that the marine extinctions may not be as profound as has been thought. Another school of scientists point to the fact that the Deccan Traps in north west India (formed from huge eruptions of basalt lava) coincide with the Cretaceous—Tertiary boundary. Could this have been an alternative source of the iridium anomalies and provide a more domestic explanation for the extinctions? After all, we know that volcanic eruptions have had profound climatic effects even within historical times. The catastrophists have countered that the type of volcanic eruptions which produced the Deccan Traps (continental flood basalts) are not of the explosive kind usually associated with gross climatic effects. They also claim to have found the site of the meteorite impact: Chixulub on the Yucatan Peninsula in Mexico, the ancient home of Mayan civilisation. There have even been claims that

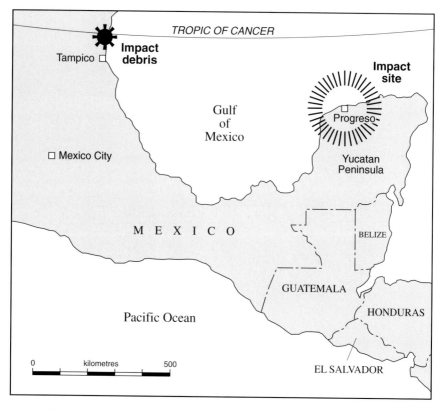

above: Recent research suggests that an asteroid may have landed in the sea just off the tip of the Yucatan peninsula, Mexico, about 65 Ma ago.

evaporation of sulphur compounds at this site may have contributed to acid rain at the crucial interval. Although the jury is still out on the details, it does look as if the meteorite catastrophists are carrying the day. But palaeontology is full of surprises and it is not inconceivable that a new fact will turn the table once again.

This short account of the Cretaceous—Tertiary extinctions shows just how complicated it can be to work out the causes of one of the most important events in the history of the biological world. It is obviously worthwhile finding an answer, sifting through all the contradictory evidence to find the real causes, even if the 'right' answer always seems to elude us. The whole story could be repeated for the Permian—Triassic mass extinction. There does not seem to have been a meteorite impact at the end of the Permian, but at this time there was a period of marine regression, when the separate fragments of Pangaea finally collided to form a supercontinent. This continental aggregation resulted in a climatic change to an extremely arid environment in many parts of she world. It is possible to combine these effects together in various ways to explain why so many animals became extinct at about the same time, although the late Permian event may have been more leisurely than its late Cretaceous counterpart. Whatever the final explanations for the mass extinctions turn out to be, it is undeniable that their effects have been the most important thresholds through which the fauna and flora passed to make the modern world.

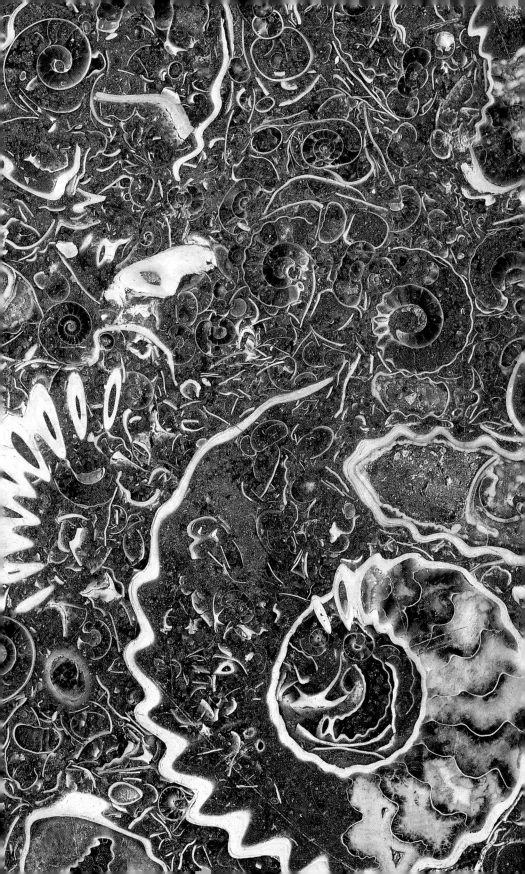

Chapter Seven

FOSSILS IN THE SERVICE OF HUMANS

· ·

EVERY time you drive a car you are able to do so because of the photosynthetic activity of plants many millions of years ago. The growth of industry to the level it reached in the 19th century was related to the exploitation of coal, just as the world economy is today related to the extraction of oil from rocks. Most of the items that we take for granted, from plastic spoons to television sets, ultimately owe their existence to energy derived from consuming fossil fuels. In times of inflated oil prices this dependence becomes manifest in price rises in all sorts of other commodities that seem at first glance to have nothing to do with oil. It would not be overstating the case to say that western society owes its present affluence to fossil fuels. As each year passes the resources dwindle alarmingly, and it has become a cliché of modern times that the process cannot go on indefinitely. Humans are exploiting the past, plundering the fossil record, and this can only be done once.

opposite: Ammonite marble: surface of a specimen containing *Asteroceras* (large shells) and *Promicroceras* (small shells) from the Lower Lias, Somerset, UK.

So far we have looked at the ways that fossils have been used to interpret the history of the Earth, and also as items of interest in their own right. It is appropriate now to take a look at the ways in which fossils are of practical use, directly or indirectly. Since the initial impetus which stimulated people to look more closely at fossils was economic (notably the need to produce better geological maps) there has been a constant interplay between the academic side of palaeontology and the ways palaeontological results can be used for industrial means. For such industrial purposes it is neither here nor there to know how trilobites or dinosaurs lived, or the problems of classification which exercise the minds of many palaeontologists.

MICROPALAEONTOLOGY

Generally speaking the abundance of fossils is inversely proportional to their size; the tiny ones being the commonest. The chances of finding an identifiable dinosaur or fish from a borehole are very slim indeed, and so the kinds of animals that dominate the landscape in reconstructions of the past do not figure prominently in the records of oil companies. The study of microscopic fossils — micropalaeontology — has become more and more important over the last half century. Not only can microscopic fossils be recovered from boreholes with a diameter of a few centimetres, but they can also be teased out of rocks otherwise devoid of organic remains. Some types of microfossil seem to have evolved almost as fast as their larger contemporaries, so they can be used in just the same way, as clocks to measure the passage of geological time. It is their usefulness in stratigraphy which gives them their industrial importance. To correlate between one borehole and the next, or between a whole series of boreholes and those from a different country, one of the simplest and cheapest methods is to identify the fossils. Some of the fossils that are used are introduced in the following sections of this chapter. Micropalaeontologists often become specialists in one group of microfossils of a particular geological period. Now that microfossils have even been found in the later Precambrian rocks, their use spans much of the geological column. Experts on tiny fossils are employed by geological surveys, as well as oil and mineral prospecting companies.

CONODONTS

Conodonts are tiny, tooth-like fossils 1 mm (0.04 in) or so long, made of calcium phosphate. They can be abundant fossils, occurring in thousands from 1 kg (2.21 lb) of rock. They are used as important stratigraphic indicators in rocks ranging in age from Cambrian to Triassic, when the conodonts apparently died out. In life, individual conodonts occurred together in clusters consisting of opposing pairs of identical conodonts, and several different kinds of conodonts often went into these 'apparatuses'. Conodonts are of special use in dating limestones. Because of their phosphatic composition they do not dissolve in acetic acid, so if limestones are put into a bath of acetic acid all the calcium carbonate dissolves leaving behind a residue of insoluble products, including all the conodonts. You do not have to be able to see conodonts on the surface of a piece of limestone to be able to find them in solution. Hundreds of different conodonts have now been described.

above: Conodonts from different geological periods: (top) Devonian of the Timan-Pechora region, Russia about 360 Ma old (magnification x 4.4); (middle) Silurian of the Ludlow area, Shropshire, UK about 420 Ma old (magnification x 3.8); (bottom) Ordovician of northern Estonia, about 465 Ma old (magnification x 4.2).

The conodont animal was discovered in the 1980s, and as so often occurs in palaeontology it happened almost by chance. Two palaeontologists were studying Carboniferous arthropod fossils from the Scottish Carboniferous, and conodonts were the last thing on their minds. A curious, worm-like fossil a few centimetres long came to their attention. It was not very conspicuous, which is presumably why it had escaped attention before. Only later did it

left:
Reconstructions of the conodont animal, representing several species. Note the eyes at the front end.

become apparent that there were conodonts within this creature, and not only did the conodonts belong to a known variety, they were also associated together in a natural apparatus. The chance of this being a fortuitous association was remote. Subsequently, more specimens have come to light, which confirms the correctness of this judgement. The conodont apparatus occupied only a fraction of the length of the whole animal. The animal that bore conodonts was a peculiar one. The long, slim body turned out to be no worm at all, but almost certainly a chordate — a member of the group which includes the vertebrates. The conodont animals were a successful side branch, which apparently left no descendants.

OSTRACODS

It is common to find bedding planes of limestones or shales covered with tiny oval blobs looking like diminutive beans. These are the shells of ostracods, a group of small crustaceans which are encased in a pair of shells, looking rather like miniature clams. Like all other arthropods they have paired appendages, which they use for feeding and for moving about, and the superficial resemblance to molluscs does not indicate any zoological relationship. Ostracods are remarkable for having proportionately the largest penis in nature. Most ostracods are small enough to be classed as microfossils, but a few approach the size of a broad bean, and these species can be as conspicuous as brachiopods when they are broken out of the rock. The smaller species, 1–2 mm (0.04–0.08 in) across, frequently occur in huge numbers, and under the microscope show a beautiful variety of fine detail, which makes them excellent guide fossils. Ostracods occur in both marine and fresh water environments (different species of course), and there are specialized species adapted to brackish water conditions as well, so that they are also useful indicators of past sedimentary conditions. The tiny shells are composed of calcium carbonate. The ostracods have a history possibly extending back to the Cambrian, and they continue as a varied and successful group today. They have been widely used in dating rocks of Mesozoic age, but there seems

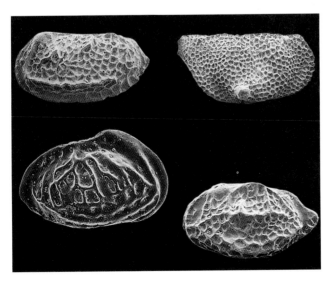

left: Ostracods: small arthropods with two valves. These examples are Jurassic in age.

opposite top: Nummulitic limestone made up of the hard parts of numerous foraminifera.

opposite bottom: *Nummulites gizehensis* embedded in Eocene limestone from 2 km (1.2 miles) north-east of Gizeh, west of Cairo. These specimens have been sectioned naturally by the break in the rock to reveal their internal chamber arrangement (about 7 cm or 2.5 in across).

every reason to suppose that they will be used more widely in dating Palaeozoic rocks as well. The electron microscope has given a new dimension to their study, because it enables even the finest details to be seen at high magnification. Although some ostracods, particularly fresh water ones, are smooth and featureless, others are covered with pimples and ridges that make them highly distinctive fossils.

FORAMINIFERA

Most single-celled foraminifera are very small, less than 1 mm (0.04 in) in diameter, and must therefore be studied under the microscope. Many of the Tertiary species are found in unconsolidated sediments, and can be extracted from the rocks by sieving; the 'forams' can then be picked out by eye under a microscope. These well-preserved fossils are studied using an electron microscope to make high magnification pictures (micrographs). In spite of their small size many foraminifera show a wealth of fine detail; some of them are covered with tiny thorns, and many have long spines, or strange, lip-like structures. Their general shape varies from species that look like tiny necklaces, to others that resemble a small bunch of grapes, and yet others with inflated, globular chambers. *Nummulites* were abundant in some earlier Tertiary rocks. They resemble small coins in size and shape, but inside are composed of numerous whorled chambers. They make up the limestones from which the celebrated Pyramids of Gizeh, Egypt were constructed.

Forams are a tribute to the flexibility of the single cell, and this flexibility means that they are among the most useful of geological clocks for determining the age of sedimentary rocks. This is especially true of the planktonic species, which are so numerous in the Mesozoic and Tertiary. Oil companies employ numbers of 'foram men' to correlate the rocks in which oil is found, and some of these experts have become renowned authorities on the history of the group. Since drilling on the deep sea floor has become a practical possibility, foraminifera have acquired an added importance. Changes in the populations of planktonic species accurately reflect the changes in climate in the more recent geological past, and they can be used to read the ages at which particular pieces of the ocean floor were created at the mid-ocean ridges. Of all fossil animals it is the tiny foraminifera which have proved most useful in the revolution of geological ideas that accompanied the theory of plate tectonics. Forams have to be used carefully though, because some of the shapes that evolved in the planktonic species were produced independently at different times from different ancestors. Evolution played the same game more than once to produce superficially similar end products. Often it is necessary to look inside the minute chambers to find out the most intimate details of their construction. Many of the older occurrences are in hard limestones from which it is not possible to extract the entire animals, and here thin sections are the only means of studying the evolution of the group. Forams have been used extensively to correlate rocks of Carboniferous and younger age, and really come into their own in the Tertiary, after the disappearance of the ammonites in the late Cretaceous. Their ubiquity even led one worker to suppose that all rocks (even including igneous ones!) were made of foraminifera — an impression which might be forgiven after a hard day sorting out hundreds of the animals from residues.

COCCOLITHS

Even smaller fossils are used to date rocks of Mesozoic and younger ages. Among these, one of the most important groups are coccoliths, minute plates a fraction of the size of a foraminiferan (usually with diameters of only a few μm). In life they were the covers for resting cells of single-celled algae (coccolithophorids), with many coccoliths to a single cell. Some of them are so small that they lie at the limit of resolution for a light microscope, and again their study has been revolutionized by the use of the electron microscope. Coccoliths form beautiful rosettes of calcite plates, every species with the plates stacked in a different fashion. Although so small, they seem to have changed rapidly through time, and are very useful in dating rocks from boreholes and similar small samples. They have their use in criminology, as they can be used to identify the source of even the merest smear of sediment on a suspect's shoe. They can form a large part of the fine fraction of sedimentary rocks, as in the soft, white Cretaceous chalk. Coccolithophorids are very important today in controlling the nutrient balance of the world's oceans.

MICROSCOPIC PLANT REMAINS

The fossils of spores and pollen are extremely small, but they are surprisingly tough. The walls of these tiny grains are composed of a very resistant organic material, that serves in life to contain the vital reproductive material, and has a very high chance of becoming fossilized. So insoluble are the walls of pollen grains that they even resist attack by hydrofluoric acid, possibly the most unpleasant and voracious acid there is. The tiny fossils survive after the rocks that contain them are broken down by acids and other chemicals, and it is not uncommon to find these microfossils in rocks that otherwise lack all trace of organic material. This is because pollen grains are wind-borne, and may come to their final rest in sediments from environments in which other life is lacking. Although small, different spores and pollens have distinct peculiarities which allow the recognition of different species under the microscope. For example, they may be ornamented with ridges or spikes, inflated or compressed, and have many different shapes. They are widely used in dating rocks, and are of prime importance in the dating of fresh water or terrestrial sediments in

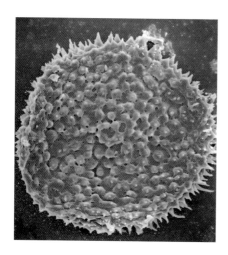 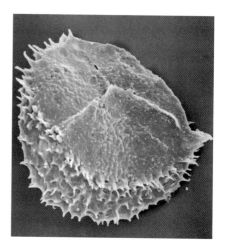

above: Spores from a Middle Devonian lycopod. Although widely dispersed and useful for correlating rocks around the world, their special interest is that they were found in situ in their parent plant. Left, viewed from the outside; right, surface were other spores were attached.

which the kind of guide fossils, which are typical of marine sediments, are absent. So important has the study of spores and pollen become that it even has its own scientific discipline — palynology — and palynologists comprise a significant fraction of the palaeontologists employed by oil and mining companies. The identification of spores is vital to the understanding of both the formation and dating of coal deposits. Because the kinds of plants that produce spores and pollen are Devonian and younger in age, the use of these fossils is confined to the later Palaeozoic, Mesozoic and Tertiary.

A particular use for palynology is in the dating of climatic fluctuations during the Pleistocene Ice Ages. The suite of pollen varieties in these relatively recent rocks can be correlated with species still living, so that cold phases will be associated with a predominance of Arctic species, and warm, interglacial periods will be reflected in the appearance of the pollen of subtropical species of plants.

Similar spiky, spherical fossils can also be recovered from marine sediments. These are the cases of resting cells of marine algae (hystrichospheres). They too have become the subject of scientific attention, and palynologists can now contribute much to the dating of marine rocks. Often these tiny fossils can be recovered by treating shales which have no other fossils. A related group of fossils, the acritarchs (see Chapter 5) can be recovered from early Palaeozoic rocks, and even some Precambrian rocks. The future of industrial palaeontology will probably be closely linked with these unassuming, tiny spheres, which can be recovered from the rocks in their thousands, and can solve the problem of dating rocks from which other fossils cannot be recovered.

COAL AND OIL

In 1769 James Watt patented the design of his steam engine, and so began the plunder of fossil fuels. The use of coal as a cheap, accessible source of energy fuelled the growth of industry, and gave the ascendancy in world economics to the industrial nations which lasts to the present day. The world of trade can be viewed as a vast machine, one which turns on the consumption of energy, mined from the geological record. In this century oil became an even more important source of the energy needed by the machine, which was growing inexorably as population increased, and the material expectations of the workers grew with it.

As the complexity and extent of the industrial process has expanded to its present gross proportions, a few people have started to question whether the whole machine might grind to a halt. There will be an end to the reserves of coal and oil, they cannot grow again in the rocks once they have been removed. We have to hope that human ingenuity will be equal to the challenge of finding replacements for fossil fuels. Radioactive minerals and solar energy may go some way towards filling the gap, but new sources of energy will also have to be discovered.

When we burn coal or oil we are releasing the energy that the Sun beamed upon the Earth for hundreds of millions of years. All the energy stored in the ground depended on the photosynthetic activity of plants, whether huge trees in the Carboniferous period or minute, planktonic algae. Coal is the black, carbonized, compressed concentrate of plants, especially trees, and only forms under certain conditions; it is not sufficient to have a stand of trees that gets swamped by sediment. To produce a workable seam of coal generations of trees are needed. Under normal conditions wood decays, its organic material is consumed by bacteria and fungi until it is friable and porous, and ready to fall back into the soil from whence it came. For trees to turn into coal they have to be protected from this normal, aerobic decay. The right conditions for this to happen pertain in swampy environments, where large areas may be fetid with lack of oxygen a few centimetres below the sediment surface. In humid, warm forests like many of those of the Carboniferous, plant growth would have been rapid, but even so it would take several hundred thousand years to accumulate enough material to form a coal seam. It is also essential for the whole area in which the coal is forming to be slowly subsiding, so that in time the tree trunks became part of a thick sequence of rocks. Over time, the sea would have made periodic advances across the coal swamp, causing a retreat of the trees and burying the potential coal beneath a blanket of marine sediment. At other times increased floods of sediment from rivers would dump their sands and silts. As burial of the coal proceeded volatile materials were driven off. It requires a great deal of burial (3000 m, 9843 ft or so) before the process is advanced enough to produce a coal of much utility. Very deep burial, or the heating effects of nearby igneous activity, are necessary to produce the highest quality, nutty coal known as anthracite. All this takes a lot of time, and most of the higher quality coals are correspondingly of Palaeozoic age. Mesozoic or Tertiary coals are often known as lignites — fossil peats in which the process of coalification has not proceeded to the full degree. Carboniferous coals accumulated in a number of separate basins, and the spores of the

left: Fossil horsetail, *Annularia sphenophylloides*, from the Carboniferous. This specimen is preserved as a carbonaceous compression on a very fine-grained and well bedded sandstone, a preservation common in Carboniferous rocks. The little circlets of flat 'leaves' at regular intervals on the jointed stems serve to distinguish this plant from others in the Carboniferous coal-shales (8 cm or 3 in long).

right: Carboniferous seed-fern plant *Neuropteris*.

coal measure plants are useful for correlating rocks from different basins. Some coals are almost entirely composed of masses of such spores. The subsequent history of the coal basins is complicated. They were nearly always fractured by faults, breaking the coal seams up, and so contributing to the dangers of mining which are part of folklore. All coals reveal their plant origin in the impressions of bits of bark, or occasionally leaves, that can be seen on the shiny surfaces.

Oil, the dark fluid that comes gushing from deep wells to stock the refineries of the world, is just as much a product of biological activity as is coal. The ultimate origin of oil is the organic material contained in many sedimentary rocks, including marine mudstones and shales, especially those that accumulated under stagnant conditions. The fixing of the Sun's energy in the sea is brought about by photosynthetic plants, particularly single-celled algal plankton. This is the ultimate foundation of the whole economy of the sea, and so the raw material for oil is in a sense the produce of fossil sunshine in the same way as coal. Oilfields have been found in rocks as old as Precambrian. The hydrocarbon compounds typical of oil are released from organic material in sediments by bacterial activity, but at an early stage are dispersed throughout the rock. Time and burial are essential to produce a workable oilfield. During burial and compaction oil is squeezed out of the source rock, together with pore-water, and can then migrate to sites where it becomes concentrated. These are the reservoir rocks, often rocks which would initially have contained no oil at all, but serve to store it because they are highly porous, concentrating all the dispersed hydrocarbons into an economic pool. Finally it is necessary to trap the oil, to prevent it from migrating to the surface. Where oil does reach the surface it produces natural seeps of oil, the most famous of which are asphalt 'lakes' like those in Trinidad, and it is probable that many of the first wells were drilled in the vicinity of such obvious surface shows.

The search for oil has led to wider and wider exploration. At first the most accessible fields, like those of Texas, were exploited. The vast resources of the Middle East, which are the main source of supply at the present time, were tapped from somewhat less hospitable territory, but now any area with sedimentary rocks is explored for its oil potential, whether high in the Arctic, or under the sea. The latter areas are naturally confined to the continental shelf, where the sedimentary cover continues to the continental slope — there is no oil in the deep sea. Some of these new areas have proved highly productive, like the British North Sea field, but sooner or later all the possible sites will have been explored, and then there will be nowhere else to drill. The coal industry may anticipate a renaissance before the end of the 21st century.

The classic oil trap is an anticlinal fold in the porous reservoir rock, capped by some impervious strata that make it impossible for the oil to migrate further and escape. The oil is often capped by a field of gas under high pressure. This is why when the drill penetrates the petroleum field the oil gushes out, and if the gas ignites this can result in spectacular, but dangerous 'blows'. Many other kinds of structures can produce important oilfields. The oil can be concentrated into porous, sandy lenticles within otherwise more shaley rocks. In some cases the porous rocks can be fossil reefs, or corals or bryozoans. Reef rocks of this kind usually have a very

high porosity because of the gaps between all the framebuilding organisms. They usually also have considerable lateral extent, and for these reasons reef rocks at depth, sealed by less permeable strata can be very important sources of oil. This explains why hard-nosed oil executives support palaeontological research into the geology of fossil reefs. In the Middle East and elsewhere oil traps have formed against the sides of salt domes. These domes of deeply buried salt are actively rising, almost as if the salt were behaving like an igneous

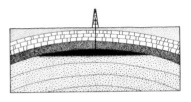

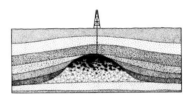

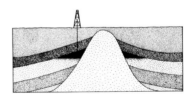

above: The geological circumstances in which one may find accumulations of oil. Top, an anticline; middle, a reservoir in fossil reef; bottom, a salt dome trap.

magma. The analogy is not altogether misplaced, because the salt does act in the manner of an igneous intrusion, flowing into the domes and pushing upwards as the load of sediment on either side pushes down. As well as sandstones and reef rocks, magnesian calcite rocks known as dolomites are often full of little cavities which may get filled

above: The surface effect of salt doming, Delicate Arch, Arches National Park, Utah, USA.

with oil under the right geological circumstances. This is one rock where the petroleum geologists and the palaeontologists might be forgiven in parting company, because dolomites of this kind are one of the poorest kinds of rocks in which to find fossils. All petroleum-bearing rocks, when outcropping at the surface, have a characteristic smell if freshly broken — rather an unpleasant one, reminiscent of greasy rags discovered in a forgotten corner of the garage.

Today, oil is recovered from greater and greater depths, and it is usually impossible to infer whether the right kind of geological structure is present there just by looking at the surface geology. A great deal of exploration is done using geophysical equipment which senses out the most important structures, and records the differences between permeable rocks like sandstones and impermeable shales and mudstones. Drilling now is nothing like the 'wildcat' operation it once was; though even today luck plays an important part in the productivity of any strike. Once drilling starts, the palaeontologists (usually micropalaeontologists) come into their own, identifying the fossils recovered and keeping a log of the age of the rocks through which the drill is passing. In some cases oil is found so consistently in rocks of a particular age it almost looks as if it were seeking out the characteristic fossil. In this circumstance it is likely that the fossil indicates a suitable facies for the formation of an oil reservoir, and here the fossil is used as something more significant than a mere calibration on the age of the borehole rock.

MINERAL DEPOSITS

Many mineral deposits are associated with igneous and metamorphic rocks, and fossils are rarely relevant to their understanding. Other minerals are often found in sedimentary rocks, and fossils are used to correlate the formations containing such minerals in just the same way as in the oil industry. Some of the commercial deposits of iron are in the form of beds of iron-bearing oolites extending over many square kilometres. These include marine deposits with a characteristic fossil fauna consistently associated with the ores. Phosphate deposits, the basis of the fertilizer industry, are also extensively developed in marine sediments. Often such phosphates are found in sites which were once near the edge of a former continent. The phosphate was probably introduced into the sediment as a result of the upwelling of deep oceanic water, which tends to occur in such locations, as it does off Peru today. Upwellings prompt a wide variety of animal life in the seas today, and are also associated with rich fossil faunas in the past. Phosphatic deposits as old as Cambrian have been found associated with a wealth of trilobites. Other peculiar fossil beds have special commercial applications. For example, the diatomaceous 'earth' of the West Indies, a rock almost entirely composed of the tiny fossil shells of diatoms, has a variety of industrial uses, including use as a special filter.

right: Fossil ammonites used as an ornamental stone, *Asteroceras marstonense* and *Promicroceras marstonensis*, Marston Magna, Somerset, UK.

ORNAMENTAL STONES

A walk around a European capital city can often be a synoptic guide to the different fossils to be found in that country, or at least to a range of those to be found in limestones. Particularly in older buildings, when builders had time and money, and labour was cheap, the floors, walls, columns and even the ceilings can be covered with a skin of ornamental stone. Many such facings contain sections of fossils. Many of course, do not, being slabs polished from igneous rocks like granite, or metamorphic rocks like marble. A great variety of limestones have been used as 'marbles' (a confusion of terminology has led all predominantly calcium carbonate facing stones to be called marble, whereas the geologist restricts the term to metamorphosed carbonate rocks) for ornamental purposes, and often the presence of fossils lends them their peculiar charm. Coralline rocks are particularly in demand, and the polished sections afford an excellent way of studying the internal structures of these animals. Other common facing stones include those largely made of the stems of crinoids, which make a bold patchwork of rods and struts in white calcite, contrasting with the darker matrix. In Scandinavia there are beautiful red limestones of Ordovician age containing the remains of straight nautiloids. In southern Europe such ancient limestones are not available, but Cretaceous rocks enclosing the peculiar bivalve molluscs known as rudists are often polished to great effect. Tertiary limestones containing masses of the giant, single-celled foraminifera *Nummulites* have also been widely used for ornamental purposes, looking like great masses of spiral nebulae adrift in a finely grained groundmass. Limestones formed by layers of algae have been used in the manufacture of ornaments. One such, which was popular in the 19th century, is the Triassic Cotham 'Marble' which displays patterns looking like a landscape of trees, picked out by the layers of sediment, which are the product of algae and bacteria. The Purbeck 'Marble' was extensively used as an ornamental stone in English cathedrals; columns of this rock surround the main pillars of regular limestone in Salisbury Cathedral. This Jurassic limestone is almost entirely made of the shells of one species of gastropod, and the polished surfaces provide sections through this fossil from every angle. The list could be extended indefinitely: wherever good, homogeneous fossil-bearing limestone is to be found it may be used as a facing stone.

DECORATIVE USE OF FOSSILS

Fossils have been used to manufacture many different kinds of decorative objects. They have been found as talismans associated with cave cultures of *Homo sapiens*. The North American Indians used the small oval trilobite *Elrathia kingii* to manufacture a necklace composed of many examples of this fossil. The same species is mined commercially today to produce everything from tie pins to paperweights. The most consistently used fossil material is amber, fossil resin, which was discussed in Chapter 1. Amber fossils include a wide range of insects and spiders. It is often shaped into drop-like pieces, and carefully matched for colour, before being mounted in necklaces, pendants, and earrings. Some of the 19th century examples of

left: The trilobite
Elrathia kingii.

amber jewellery are particularly fine. Fossil wood which has been replaced by the hard mineral silica often retains the finest details of its cellular structure. It takes a very high polish, and is an attractive deep reddish-brown colour when cut. This has been exploited in the manufacture of a variety of table ornaments, and the larger trunks have been cut through to make extremely heavy table tops, which would certainly be immune to the normal stains of domestic use, and must form the oldest tables in existence. In the Jurassic rocks of Yorkshire another kind of fossil wood is preserved as jet — a dense, very dark material that is the origin of the phrase 'jet black'. In the 19th century jet had a considerable vogue in necklaces and the like. The brilliant black colour of polished jet was adopted by Queen Victoria after the death of Prince Albert, and thence by society, as an ornament that was both decorative and decorous.

It has recently become a fashion to treat well-preserved, large fossils as objects of beauty in their own right. Fine examples of ammonites or fossil fish can be found on sale at inflated prices, described as 'Nature's Sculpture' and the like. Good looking fossils can command high prices at auction, and there is now concern, particularly up in the United States over the commercial exploitation of the biological past. The *Tyrannosaurus rex* specimen known as 'Sue' was sold for millions of dollars. The increased appreciation of the value and beauty of fossils is something to be welcomed. The only problem is that fossils, unlike living animals, cannot breed and replace themselves. Like coal and oil they are a finite resource. Classic fossil localities are easily worked out, and every palaeontologist has had the experience of arriving at a well known locality to find nothing left but a large hole. Fossils are valuable, each one in its way is a small miracle, but they are most valuable for what they can tell us about the past.

PALÆOXYRIS CARBONARIA, Schimper.
Figured by R. Kidston in Proc. Roy. Phys. Soc. Edinb., vol. ix.

COAL MEASURES.   COSELEY, near DUDLEY.
[V. 1178]   (Johnson Collection.)

Chapter Eight

MAKING A COLLECTION
· ·

MAKING collections seems to be almost an instinct with many people. A fossil collection is easy to store and maintain, and is the best way to get to know the many kinds of organisms in their various modes of preservation. Some people may prefer to make collections of fossils from the rocks in the vicinity of their own home, others may deliberately try to get a wide coverage of the different kinds of fossils from rocks of diverse ages. It is surprising how quickly a collection grows; if given a chance it soon begins to oust the collector from house and home. This chapter gives a little practical advice on how to find fossils, clean them, identify them and store them. There is no doubt that some people have an almost supernatural facility for finding fossils. Fossils seem to fall out of the rock for them, whereas others labour long and hard with no reward. There is no substitute for returning time and again to the same site and carefully examining anything that looks organic. The only two essential tools for this are a good geological hammer and a hand lens. Geological hammers come in two main varieties: those with a wooden shaft, and those in which the head is welded to a hardened steel shaft. Either is perfectly satisfactory for hammering sedimentary rocks.

Otherwise all that is needed are old newspapers (for wrapping) and patience. It is good practice to label each specimen in the field. It is amazing how your memory lets you down.

opposite: A variety of echinoderm fossils may form the basis of a specialised collection

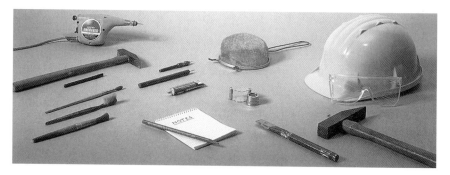

above: Implements for use in the field and for cleaning fossils: tools for extracting, brushes for cleaning, sieve for extracting from clay, hand-lens, and helmet and goggles for protection.

WHERE TO FIND FOSSILS

There are no special rules about where to find fossils. There are certain restrictions: if rocks have been heavily metamorphosed there is a slim chance of recovering good fossils, but even this is not impossible, and there are examples known where fossils have survived the most appalling maltreatment by heat and pressure. In general, good sedimentary rocks are fossil-bearing. There are a large number of known localities, that is, places where fossils have been recovered for a long time, and which are recorded in the geological and popular literature. At such places fossils are sure to be found, but it has to be remembered that generations of hunters have been there before, and that the inspiring specimens now residing in museums were probably collected 100 years before, when the ground was in prime condition. Nowadays the best collections can often be made from the same geological horizon as the classic localities, but from sites a few hundred metres to a kilometre away. There are some areas within which the rocks are thoroughly saturated with fossils; in these cases it is just a question of having enough time to gather all that may be collectable. The Jurassic rocks of the Dorset coast in southern England are one of the best known rock groups of this kind. Formation succeeds formation, all of them rich in fossils. Even if the most spectacular fossils have been removed there is no chance of the localities being 'collected out'. The same might be said of Cretaceous deposits in large areas of the interior of the United States, or the Devonian of Canada. Such areas are certainly the most encouraging in which to make a first collection. They usually represent former deposition in shallow marine habitats, rich in life, where the sediments have been little disturbed subsequently, and where the soft sedimentary rocks yield up their treasures with little resistance. Even in these well-known localities a lucky hammer blow may turn up a species new to science.

Such prolific sites apart, most sedimentary rocks do yield fossils with prolonged searching. Sandstones are usually the least rewarding. Some sandstones were deposited in desert environments and scarcity of fossils in these is hardly surprising. Others are turbidites, and fossils are rare in these too, although the presence of trace fossils should not be ignored. Limestones only occasionally lack any organic remains (apart from Precambrian limestones). If the reader is attracted by finding older fossils there is the increased possibility of structural complication in sedimentary rocks of this antiquity. Cambrian to Devonian rocks (and younger rocks in some areas) are often affected by cleavages, which make the rocks split in directions unrelated to the original bedding (in which the fossils lie). Even so, it is quite feasible to recover fossils by looking for traces of the true bedding and smashing the rock so that it breaks in approximately the right direction. This can be a heart-breaking process, trying the patience of even a professional. If the rocks are folded as well as cleaved, it is possible to find places where the cleavage and bedding coincide, and this is the premium site in which to search for fossils.

Many rock formations, although they do yield fossils, have them concentrated into small pockets in certain localities. This particularly applies to formations of fresh water or terrestrial origin, probably reflecting the patchy distribution of lakes and pools where fossils can be preserved.

The Old Red Sandstone formations of Wales and Scotland locally contain fish, eurypterids and other exciting fossils, although great stretches of the rocks can be disheartening to the casual searcher. Here persistence, and a little luck, are indispensable, and there is always a thrill in the discovery of any well-preserved fossil.

Coal measures usually yield fine remains of plants, but the best of these are not in coal itself, but in associated shales. There is always the possibility of finding one of the seams in which insects, like giant dragonflies, or vertebrates are also preserved. Rarer fossils are often concentrated in this way into particular bedding planes, and so are easily worked out by the over-zealous collector. It is essential to note the exact horizon of any unusual find, so that it can be located again.

There are many kinds of exposure in which it is possible to find fossils. The most obvious exposures are those in sea cliffs, and these also provide the best sections through different formations. In desert regions exposure can also be good, and the same is true of Arctic areas or high mountains, although collecting there is not without its attendant hazards. Never take chances climbing high or crumbling rocks on the supposition that the best fossils are to be found slightly out of reach! In domestic landscapes like that of England or the eastern United States most of the best inland exposures are in quarries or on road cuttings. It is always worth examining temporary exposures cut during road widening. Here, local knowledge is a great advantage: if you are 'on the spot' you can collect quickly before the site is backfilled and the chance lost forever. Occasionally these temporary cuts open up seams of important fossils that have never been found before. It is always necessary to ask permission to enter any working quarry; most quarry owners are quite happy to let in the foraging palaeontologist, but they do not wish to have accidents happening on their property.

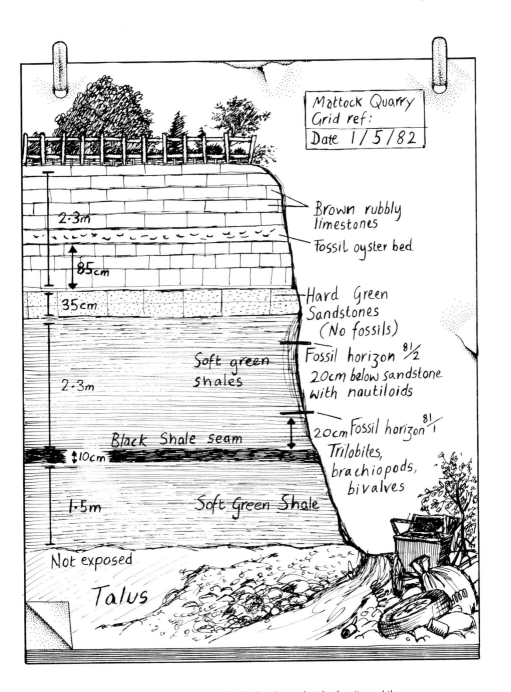

above: A page from a field note book showing a sketch of a site and the location of fossils.

opposite: Dinosaur excavation, southern Sahara, Niger, 1988. Cleaning and hardening of the bone elements prior to undercutting and the application of a plaster jacket.

In hill regions the best sections are often along the sides of streams that cut through bedrock. Permission from farmers should be obtained before trespassing on their land; otherwise you may encounter an irate farmer after unwittingly straying on to his property. On very overgrown sites it is an advantage to add a grubbing mattock to one's equipment, with which to hack out the weathered rock to get access to a clean face.

Regions well-known for their geology often have guide books published about them, by Government Surveys or by local natural history societies and museums. In the UK, the Palaeontological Association publishes guides to the fossils of some famously productive formations. Alternatively, nearly every area has a geological map, published by the National or State Survey, showing the pattern of outcrop of the rocks in the area. You should buy a map showing the solid geology rather than a drift map showing the soil types. Maps show the different rock formations, so by tracing the locality of an outcrop on to the map, it will be possible to identify the geological horizon from which any collection is made.

It is essential to note exactly where the fossils were recovered, both from the point of view of making a personal record, and as an indispensable aid in their subsequent identification. It often helps to make a sketch of the site with the horizons from which the fossils were recovered clearly labelled, preferably in metres from some distinctive marker which can be readily identified on a return visit. The different kinds of rock should also be marked, together with any peculiarity of particular beds, like the presence of trace fossils, or a change in colour. Such observations are the basis of stratigraphy, and a well-kept notebook will jog the memory when the fossils are being examined back home.

opposite: Using a vacuum powered engraving tool to remove the matrix from around a limb bone of the dinosaur *Edmontosaurus*.

CLEANING FOSSILS

There is always a temptation to chip out a sea urchin or a trilobite as completely as possible while still in the field, just to get a proper look at it. This is nearly always a mistake: a careless blow with a hammer can destroy what might, with patient cleaning, turn out to be a fine specimen. It is far better to wrap the specimen up carefully for cleaning at home. It is important to keep the other half (counterpart) of the specimens, which often show details lacking on the specimen itself. Most fossils are not so fragile as to need special treatment in the field; careful wrapping is sufficient. Some however, are very delicate and need to be toughened up before removal by using solutions of resins. If one is lucky enough to stumble across the remains of a large vertebrate, the best thing to do is to leave it exactly where it is, and contact a museum with the expertise to extract it properly. Otherwise, it is possible to destroy vital information.

Once the fossils are safely home and unwrapped, the cleaning process can begin. Some fossils tend to crack out completely, graptolites in shales for example, and little needs to be done to these. Sometimes, however, the shale still partly covers the specimen, and in this case a sharp tap with a small chisel usually suffices to break the shale along the bedding plane containing the rest of the fossil. The crucial point to remember when trying to dig out a fossil is that damage will result if the cleaning implement is much harder than the fossil and used directly against it. Fossils in shales and mudstones tend to be soft, and great care is necessary to avoid damaging them. A needle mounted in a pin vice can be used to gently remove any covering shale. This should be done by pressing obliquely on to the covering rock, and not by stabbing at the fossil itself, which nearly always results in unsightly pin pricks.

In some cases the enclosing matrix is softer than the fossil. This is the easiest case to deal with, because the enclosing rocks can usually be removed by scrubbing. Fossils from Cretaceous chalk can be cleaned using a small bristle (but not wire) brush with water. Cleaning is much more difficult if the matrix is harder, or about the same hardness as the fossil material. There is usually no choice here but to clean off the matrix gradually by hand, relying on the tendency for rock to break off around the fossil rather than through it, at the interface between the fossil and the enclosing rock. A mounted needle can be used for this (the best are the needles used for 78 rpm records, which can still be bought in junk shops), and there are various manufactured appliances that allow the needle to vibrate very fast, producing an instrument that chips away the matrix more rapidly. Cleaning fossils in this way is quite a skilled operation, and if this is being tried for the first time it is best to start on one of the more unimpressive specimens, however tempting it might be to start with the prize exhibit. Sometimes nature lends a hand with the process, and natural weathering may have already etched out a slightly harder fossil from the surrounding rock. Many of the most impressive museum specimens have been prepared naturally for exhibition in this way, but the majority of these sites have since been picked bare of such treasures, and nowadays there is usually no substitute for hard work.

In sandstone preservation particularly, but also in some limestones, the actual shell of the fossil has often been dissolved away, and what remains are internal and external moulds. The internal mould usually comes away with little trouble. The external mould should be washed clean of any dirt, which may have to be removed by gently rubbing with a toothbrush. A perfect replica of the exterior surface of the fossil may now be made by taking a cast from the external mould. Various preparations are suitable for this purpose: a plasticine squeeze can give a quick impression, but does not take up the finest detail. For this a latex rubber solution should be used, or some of the modern resin preparations, the latter having the advantage that the casts are more or less permanent. In fact, casts of this kind are every bit as good, and in some cases better, than having the fossil itself, particularly since the internal mould also gives you all the details of the internal surfaces of the fossil.

In some limestones the fossils have been silicified (replaced by silica) while the matrix consists of calcium carbonate. In this case the rock can be dissolved in dilute acetic or hydrochloric acid (with care!), and the fossils will be left behind. After repeated washing and drying a collection of fossils preserved in this way can be mounted on slides, or in small boxes. Any phosphate fossils (such as inarticulate brachiopods) are left untouched when limestones are dissolved in acetic acid, and some beautiful and surprising fossils can be recovered in this way, even when there is not much to show on the surface of the rock.

IDENTIFYING FOSSILS

While it is satisfying to make a collection of beautiful specimens of fossils for their own sake, it becomes even more so if the specimens are identified and classified. Many people find that a general collection of fossils soon begins to take up too much room, and that they prefer to specialize in one kind of fossil (such as trilobites) or in those from a particular area or age. Putting a name to a fossil may seem a rather arid exercise, and so it is by itself. But the real point is that it is not possible to identify correctly a fossil without looking at it very carefully, appreciating the fine points of its construction, and becoming familiar with a whole range of related animals or plants. Therefore to identify is partly to understand.

To identify the general kind (e.g. phylum) of animal or plant is usually a simple task. With experience it is also easy to identify the fossil to within narrow limits, i.e. to order level. There are only a relatively few major kinds of brachiopods, for example, and their general features can be mastered with a little experience. The problems start when a precise identification is required, to genus or even species. There is no easy way to do this, and sometimes even the expert in the group of fossils will have problems. In some cases a species identification may not even be possible: for example, many brachiopods are identified from their internal structures, so if your fossil does not show these it cannot be specifically identified. It is vital to know the precise formation and locality from which the fossils were recovered. Fortunately there are reference collections in museums, which have fine specimens on display, which have (one hopes) been identified by an expert. These are often arranged rock formation by rock formation, and so it is possible to home in on a series of species with which the one in hand is to be compared. It usually happens that your specimen is not exactly the same as the ones on display; the best name is then that of the closest matching species. Most fossil animal populations, like many living ones, include a certain amount of variation, so that it is unlikely that any two fossils will be precisely the same, even if they belong to a single species. Among palaeontologists there are often arguments about whether or not a population of fossils slightly different from other examples of a species is really a different species or just a local variant.

The next best thing to using a comparative collection to identify a fossil is to use books with pictures of fossils as the basis for identification. Any available books only give a selection of typical fossils from a region, but this is a good start, and because most books will show the most common fossils, you are likely to have at least some of your finds illustrated. Often, a fossil is found which cannot be closely matched with any of the species illustrated in these books. In this case the best approach, if the fossil is an invertebrate, is to obtain the relevant volume of the *Treatise on Invertebrate Palaeontology*. This rather intimidating work is a compilation of all the different kinds of fossils known at the time it was published (different years for different volumes); each volume concerns one type of fossil (e.g. brachiopods or trilobites), and contains descriptions and figures of all the different genera of the animals. It can take a while to learn how to use one of these volumes, and the pictures vary in quality. This only enables an identification of the genus of the fossil, but this is adequate for most purposes. The volumes of the *Treatise* are not likely to be stored in a local library, but they can be ordered through libraries, and they are certain to be stocked in university libraries. Finally, many museums offer an identification service. Occasionally this kind of enquiry will result in the discovery of a new species, and then the finder may be asked to donate it so that it can be subjected to proper scientific scrutiny. This underlines how important it is to record the precise details of where a fossil is found. New discoveries are

above: Identifying fossils using a guide.

quite often made by amateur collectors, particularly when they return again and again to a favourite site to find the rarer fossils in the fauna. One of the greatest excitements of palaeontology is the possibility that something entirely new will turn up; any hammer blow could be the one that breaks out the new discovery. It is this that sustains the searcher through long hours where nothing at all is discovered.

STORING A COLLECTION

As the man who made a fortune from selling 'pet rocks' in the United States shrewdly realized, fossil specimens are remarkably easy to look after. Most fossils do not deteriorate with time, and all they require is room for storage in a dry place. A few fossils look superficially attractive if they are varnished; this is not a good practice however, because the varnish obscures a lot of finer detail, and is very difficult to remove once it dries. All fossil specimens should be labelled either with a direct identification and locality, or with some sort of code which refers you to a catalogue. When a collection begins to get large, it is impossible to remember all the details about when and where a fossil was collected, and how it was identified.

Some kinds of fossils are prone to decay. This is particularly true of those preserved in iron pyrites, such as ammonites and Tertiary fruits and seeds. They acquire feathery growths of crystals as the pyrite begins to react with the atmosphere. The eventual outcome is that the fossil collapses into a heap of dust. Varnish does not greatly inhibit the process of decay. There are inert fluids on a silicone base in which fossils like these can be stored. A cheaper alternative is glycerine, but this is messy, and gradually takes up water from the atmosphere, so has to be stored in air-tight jars.

How the fossils are arranged is a matter of taste. Some people prefer to order them phylum by phylum, others according to their geological age. If the intention is to make a very detailed collection from one area or

above: Collection of sea urchin fossils properly stored in a drawer.

formation, the arrangement might be by locality. In many ways the last kind of collection is the most satisfying, because it does not take long before the commoner fossils become old friends, and can be used to trace a single geological horizon from one locality to the next. With prolonged collecting, specimens as good as any found in museums will turn up, and the collector will begin to know the faunas as well as any expert. Geological time is immensely long, and the volume of fossil bearing rock surpasses computation. There is plenty of scope for those with time and patience to make a real contribution to the knowledge of the history of life. The evidence is just waiting to be collected.

POSTSCRIPT

Although there is no shortage of fossils they can be collected only once. Having survived the vicissitudes of millions of years of burial, when they are disinterred they become the responsibility of the finder. Equally, the locality itself is a precious entrance into the past, and is to be respected. There is nothing more depressing than reaching a known locality only to find it mutilated and degraded by some irresponsible collector trying to get the biggest and best specimen without regard for those who follow. In Britain a number of sites have been declared Sites of Special Scientific Interest, which means they are somewhat protected from over-collecting. The pursuit of the perfect specimen should not be allowed to overshadow the scientific importance of the humbler, fragmentary fossils. The whole point of palaeontology is to reveal the biological history of the world, and fossils are an indispensable means to this end. Any unusual find might prove to be the clue to understanding some new aspect of this history. The evidence of past worlds has to be revered.

Further Information

FURTHER READING

At the Water's Edge, C. Zimmer.
Simon & Schuster, 1999.

Dinosaur. Eyewitness Guides 13, D. Norman and
A. Milner. Dorling Kindersley, London, 1989.

The Dinosaur Hunters, D. Cadbury.
Fourth Estate, London, 2000.

The Ecology of Fossils: An Illustrated Guide,
W. McKerrow (ed.). Duckworth, London, 1978.

Evolution, C. Zimmer. Heinemann, Oxford,
2002.

The Evolution Revolution, K. McNamara and J.
Long. John Wiley & Sons, London, 1998.

The First Fossil Hunters, A. Mayor and P.
Dodson. Princeton University Press, 2001.

Fossil. Eyewitness Guides 19, P. Taylor. Dorling
Kindersley, London, 1989.

From the Beginning, K. Edwards and B. Rosen.
The Natural History Museum, London, 2000.

Graptolites, Writing in the Rocks, D. Palmer and
B. Rickards (eds.). Boydell Press, Suffolk, 1991.

In Search of the Neanderthals, C. Stringer and
C. Gamble. Thames and Hudson, London,
1994.

*The Illustrated Encyclopedia of Dinosaurs and
Pterosaurs: an original and compelling insight
into Life in the Dinosaur kingdom*, D. Norman
and P. Wellnhofer. Salamander Books,
London, 2000.

*Life; an unauthorized biography. A natural
history of the first four thousand million years of
life on Earth*, Richard Fortey. Harper Collins,
London, 1997.

The Meaning of Fossils, M. Rudwick. University
of Chicago Press, 1972.

*National Audubon Society Field Guide to North
American Fossils*, I. Thompson. Knopf, 1982.

The Natural History Museum Book of Dinosaurs,
2nd edn., T. Gardom and A. Milner. Carlton
Books, London, 2001.

Scenes from Deep Time, M. Rudwick. University
of Chicago Press, 1992.

Trilobite! Eyewitness to Evolution, R. Fortey.
HarperCollins, London, 2000.

Trilobites, H. Whittington. Boydell Press,
Suffolk, 1992.

The Triumph of Evolution, N. Eldredge. W.H.
Freeman, London, 2000.

Websites

NB. Website addresses are subject to change.

Australia Online Museum
http://www.amonline.net.au/earth_sciences/palaeontology.htm

Buena Vista Museum of Natural History,
Kern County, California
http://www.sharktoothhill.com/

Carleton University, Canada
http://superior.carleton.ca/~tpatters/Museum/hvpmdoor.html

The Field Museum of Natural History
http://www.fieldmuseum.org/

Museo Nationale de Ciencias Naturales,
Madrid
http://www.museociencias.com/english/index1.html

Museum of Paleontology, University of Michigan
http://www.ummp.lsa.umich.edu/index1.html

National Museum of Natural History,
Smithsonian Institution
http://www.nmnh.si.edu/paleo/

The Natural History Museum, London
http://www.nhm.ac.uk/

The Palaeontological Association
http://www.paleosoc.org/

PaleoNet
http://www.nhm.ac.uk/hosted_sites/paleonet/PalAss/index.html

Paleontological Research Institution
http://www.priweb.org/

Pella Museum, Jordan
http://www.pellamuseum.org/

Plant Palaeobiology Group, Royal Holloway,
University of London
http://www.gl.rhbnc.ac.uk/palaeo/palaeo.html

The Royal Tyrrell Museum
http://www.tyrrellmuseum.com/

Glossary
· · · · · · · · · · ·

Abyssal At great depth, off the edge of the continental shelf.
Appendages The limbs, gills and antennae of arthropods.
Aragonite Form of calcium carbonate employed in the construction of shells by some marine animals.
Astogeny Growth of a colonial organism by addition of individuals of the colony.

Bedding plane Plane parallel to the former sea floor (or freshwater equivalent).
Benthic (or benthonic) Describing organisms that live (or lived) on the sea bottom.

Calcite The common mineral form of calcium carbonate, of which many fossils are made.
Chert A hard sedimentary rock composed of fine-grained silica.
Cleavage Tendency for metamorphic rocks (e.g. slates) to break at an angle to the bedding plane (hence, plane of cleavage).
Conglomerate Coarse, pebbly sedimentary rock, such as the deposit of an ancient beach.
Correlation The process of establishing the time-equivalence (or otherwise) of sequences of rocks.

Epicontinental Surrounding the edges of continents; especially of shallow seas.
Era Major division of geological time: Proterozoic, Paleozoic, Mesozoic and Cenozoic (Tertiary and Quaternary).
Exoskeleton The exterior skeleton of the arthropods.

Facies Rock type (or collection of rock types) representing a particular environment where the rocks were deposited. (e.g. reef facies, lagoonal facies); also applied to faunas reflecting the same environment.
Fauna An assemblage of fossil animals from one site or age (a flora is the botanical equivalent).
Filter-feeder Animal that lives by filtering out small particles (usually plankton) from the sea, or fresh water.

Helical coiling The upward-spiral kind of coiling typical of many snails and a few ammonites.
Homeomorph An animal that resembles another, possibly because of a similar mode of life, but is not really biologically related.

Igneous rocks that have formed from the cooling of hot, liquid magma, ultimately derived from deep in the Earth, and of course without fossils.
Internal mould Natural cast in sediment of the inside of a shell or other fossil.
Intrusion Mass of igneous rock, which intrudes into the surrounding strata.

Living fossils Term applied to animals or plants that have survived for a long time, or at least from a time when there were many more of their kind.
Marine transgression Invasion of the sea over land area, caused by relative rise in sea level.
Metamorphic Rocks that have been altered by heat and/or pressure, usually because of deep burial or involvement in mountain-building episodes.

Oolite Sedimentary rock (usually limestone) composed of spherical ooliths, and usually formed in shallow water.

Pelagic Free swimming in the oceans.
Phyla (singular, phylum) Major zoological unit of classification, indicating broadly related animal groups.
Planktonic (or planktic) Passively floating in the oceans (noun, plankton).

Radiometric age Age given in years using the natural 'clock' of radioactive minerals.
Regression Draining of the sea from a continental area; the opposite of transgression.

Shield areas Large areas composed of Precambrian rocks that have acted as coherent, stable blocks for hundreds of millions of years.
Silica One of the most abundant compounds in nature, silicon dioxide, of which quartz and chalcedony are two of the commonest forms, and which is used as a skeletal material by a few organisms.
Stratigraphy That part of geology concerned with the description of the relationships of rocks in the field, and their correlation.
Subduction zone Area where oceanic crust is plunging downwards beneath an adjacent continental block.
Suture line Line marking the junction of the chamber wall with the shell in ammonites and nautiloids.

Test The 'shells' of echinoderms or foraminiferans.
Turbidite Sedimentary rock formed by the action of a turbidity current.

Unconformity Break between two sequences of rocks; the lower sequence is often tilted, uplifted and eroded before the deposition of the overlying one (angular unconformity).

Zone Basic unit for correlation, a unit usually typified by a characteristic assemblage of fossils belonging to one or more groups of organisms.
Zooid Individual animal in a colonial animal – applied particularly to bryozoans and graptolites.

Index
· · · · · · · ·

Page numbers in *italics* refer to
captions, in **bold** to colour plates.

Aberystwyth Grits, UK *67*
Acanthometra 74
Acer trilobatum **174**
acritarchs *87*, *129*, *197*
Acrocoelites subtenuis **58**
Actinocyathus floriformis 79
Adriosaurus suessi **167**
Aeger sp. **164**
algae *127*, 129-30
    calcareous *78*, *80*, *137*
    planktonic *85*, *132*, *198*, *200*
amber 205-6
    insects in *14*, *15*, *183*
ammonite marble *187*
ammonites *40*, **56**, **57**, *87*, *204*
    evolutionary changes *149*
    extinctions *88*, *122*, *160*, *178*
    life habits *86*
    snake's heads *7*
    as zonal fossils *30*, *33*, *34*
annelid worms *132*
Annularia sphenophylloides *199*
Anthrophorites titania **161**
Antiquatonia hindi *52*
Apatosaurus 99
aphids **163**
aragonite *73*
Araucarioxylon arizonicum **173**
Archaea *119*, *120*
archaeocyathids *80*, *82*, *137*
Archaeopteris hibernica **173**
Archaeopteryx 13-14, *102*, *103*
Archastropecten cotteswoldiae **60**
Archimedes sublaxus **165**
Architectonica millegranosa **54**

archosaurs *177*
Ardipithecus ramidus *152*
Asteroceras *187*
    A. marstonense *204*
Australopithecus *152*, *154*, *158*, *159*
    A. afarensis *153*
    A. africanus *153*

bacteria *119*, *120*, *123*, *130*, *131*, *133*
    cyanobacteria *124*, 126-8
Bangiomorpha pubescens *129*
Banwell Bone Cavern, UK *76*
Baryonyx walkeri *7*
basalt *39*, *183*
Beard, William *76*
bees **161**
belemnites **58**
Belemnotheutis antiquus **58**
birds, fossil *see* feathered
    dinosaurs
Bison priscus *76*
Bitter Springs Chert, Australia *128*
bivalve molluscs *19*, *82*
    life habits 110-12
    trigoniid *54*
blastoids *177*, *178*
borings
    bivalve molluscs *19*
brachiopods *11*, **52**, *80*, *82*, *160*
    extinction *177*
    life habits 107-10
    as zonal fossils *30*, *33*, *34*
Brachiosaurus 99
brittle-stars **61**
Broken Hill man *154*
Brooksella *131*
bryozoa *80*, **165**, *177*
Burgess Shale, Canada 11-12, *113*

burrowing *19*, *133*, *134*
    bivalve molluscs *110*, *111*
    polychaete worms *91*
    sea urchins *149*

Calanus glacialis 77
Camarotoechia sp. **52**
Cambrian period *29*, *30*
    extinctions *160*
    fossils *12*, *17*, *34*, *45*, *84*, *86*,
        134-41, **165**
    rocks *11*, *18*, *28*, *80*, *113*, *211*
Carboniferous period
    forests *177*, *198*
    fossils *52*, *79*, *112*, **163**, *165*,
        172, *175*, *189*, *199*
    glacial activity *45*, *46*
    rocks *28*, *29*, *73*
casts of fossils 218-19
Cenoceras pseudolineatus **57**
Cenozoic era *30*, *31*
Ceratolithoides aculeus *88*
Chaceon peruvianus **162**
chalk *31*, *88*, *150*, *218*
chemical fossils *24*
cherts *84*, *85*, *105*, *122*, *124*, *128*
    Bitter Springs *128*
    Rhynie *10*
chitons *138*
Chixulub, Mexico 183-4
Chlamys *112*
cladograms *151*
clams *87*
clays, red *74*
cleaning fossils 216-19
cleavage *21*
climate changes/fluctuations
    44-8, *65*, *76*

pollen dating 196
coal 197
    formation 198
    plant fossils 199, 212
coccolithophorids 88, 195
coccoliths 87, 88, 194-5
Coeloptychium agaricoides **49**
collecting fossils
    cleaning 216-19
    finding 210-14
    identification 219-22
    site recording *213*, *214*
    storage 222-3
Compsognathus **169**
Conchoecia imbricate 77
Confuciusornis sanctus 93, *103*
conodonts 189-90
continental drift 37-42
corals 79, 83
    extinctions 177
    fossils **50**, 79
    reefs 44, 78-82, 201
counterparts of fossils 8, 10, 216
crabs **162**, 177
    horseshoe 13, **161**
Cretaceous period 30, 31, 32
    climate 48
    extinctions 34, 122, 160, 178
    fossils **49**, **54**, **62**, 81, 87, 88,
    93, **167**, **168**, **169**, 218
    rocks 211
crickets 15
crinoids (sea lilies) **59**, 177, 204
crocodiles 177, 178, 180
crust, oceanic
    generation and destruction *37*
Cruziana semiplicata 18
Cryptozoon 124
crystoids **60**
Cyathophyllum sp. **50**
cycads 177
Cystisoma 97

Dactylioceras 7
Dalmanites myops **63**
Darwin, Charles 79, 144, 151
dating, rock
    radiometric 34-6, 46
    zonal fossils 32-4, 189, 191,

193, 194, 195, 197
Deccan Traps, India 183
decorative use of fossils 205-7
deep sea deposits 82-90
Delicate Arch, USA *202*
Deltoblastus jonkeri 178
desert areas 69
Devil's toenails *see* oysters
Devonian period 29, 30, 31
    extinctions 86
    fossils *13*, 34, **49**, **50**, **64**, 80,
    104, **173**, *189*, 196,
    rocks 10, 12, 28, 211
diatoms 203
Dichograptus octobrachiatus **51**
Dickinsonia 132
Dictyonema retiforme 8
Didymograptus 149
    D. murchisoni **51**, 77
dinosaurs 99-103, **168**, **169**
    excavations of *7*, *23*, *214*
    extinction 178
    theories 122, 179-82
    tracks *17*, *99*, *100*
    see also feathered dinosaurs
Diplodocus 101
Diprotodon **171**
dolomite 201-2
dragonflies 13, **162**

Earth
    age 27-8, 117
    atmosphere 117-19
    continental drift 37-42, 46
    early life 117, 119-34
    formation 117-18
    water 118
earth movements
    effects on fossils 20-1
earthquakes 38, 76
Echinoconchus punctatus **52**
echinoderms 138
    collections 209
Ediacarian era *see* Vendian era
Edmontosaurus 216
    E. regalis **169**
Elrathia kingii 205, *206*
Emiliana huxleyi 88
Encrinus liliiformis **59**

endemism 40-1
Eocene epoch 30, 31
    fossils **53**, **55**, **163**, **166**, *193*
equipment, fossil collection
    see tools
eukaryotes 123
Euparkeria 101
Eurypterus lacustris **64**
evolution 143
    early life 122
    evolutionary bursts 141
    fossil animals 148-51
    and fossil record 145-8
    humans 151-9
excavations of dinosaurs *7*, *23*, *214*
extinctions 143-4, 147
    environmental causes 48
    major events 159-60, 177-85
    viral epidemic theory 122

facies fauna 69
Favites pentagona 79
feathered dinosaurs 13-14, 93, 102,
    103
ferns 181
finding fossils 210-14
fish **166**, **167**
Florosphaera profunda 88
flowering plants 182
footprints, dinosaur *see* tracks
foraminifera 74, 88, 148, 178,
    193-4
    evolutionary changes 148-9
    life habits 87
    marine depth stratification
    90, 91
    as zonal fossils 30, 34
fossil animals
    evolution 148-51
    extinctions 144-8
    reconstructions 94-5
fossil collecting 219-24
fossil fuels 187
    see also coal *and* oil
fossilization 7
    flattening 8
    mineralization 10
fossils
    accumulation sites 69

counterparts 8, 10, 216
decay 222
decorative use of 205-7
and earth movements 20-1
enigmas 112-15, 136
fragments 21-5
hard parts 135, 136, 139, 140
new discoveries 221-2
oldest fossils 123
soft parts 11-12
use in stratigraphy 32-3, 91,
189, 191, 193, 194, 195, 197
see also chemical fossils;
microfossils and trace fossils
frogs 170

gastropods 55, 56, 138
as zonal fossils 30
geological circumstances, special
11-17
geological maps 214
geological time 27-34
and marine deposits 44
timescale 30
glaciation 45-8, 140
glacial deposits 75
Globigerina 74
Globigerinella
G. aequilateralis 90
G. calida 90
Globigerinoides 74, 148
G. ruber 90
Globorotalia crassaformis 90
Glossopteris 39
Grand Canyon, USA 43
graptolites 8, 51, 84-5, 86, 89, 216
colonies 103-6
evolutionary changes 149
extinctions 160
as zonal fossils 30, 32, 34
Gryphaea arcuata 54

Halkieria 137
Hallucigenia 113, 115
H. sparsa 114
Harpagodes wrightii 56
harvestmen 15
Heliconema 128
helicoplacoids 138

Helicoplacus everndeni 138
Hemicidaris intermedia 61
Hiatella 19
Hippurites socialis 81
Homo
H. erectus 154-5, 158
H. habilis 153, 154, 157
H. heidelbergensis 157
H. neanderthalensis 155, 157, 159
H. rudolfensis 153, 154, 157
H. sapiens 143, 155, 157, 159, 205
horsetails 199
humans
fossil remains 151-9
Hunsrück Shale, Germany 12
Hutton, James 65
Hydnoceras tuberosum 49
hydrothermal vents 119
Hypotodus robusta 166
Hypsilophodon foxii 168
hystrichospheres 197

Ice Ages 14, 44, 77, 155
ichnofossils see trace fossils
ichthyosaurs 177
Ichthyosaurus acutirestris 167
identifying fossils 219-22
Iguanodon 22
industrial micropalaeontology
188, 194
Inoceramus 150
insects 182
in amber 14, 15, 17
iridium anomalies 181, 183
iron ore, banded 127
iron pyrites 222
Isograptus 149

Janospira 113
jellyfish 131-2
jet 206
Jurassic period 29, 30, 31, 81, 160
fossils 7, 13, 54, 56-9, 60, 61,
62, 102, 162, 164, 167, 169
rocks 73, 83, 205, 210
trackways 17

Kelvin, Lord 28
Kimbe Bay, Papua New Guinea 83

Kosmoceras acutistraitum 56

Late Cretaceous extinction event
178-85
Late Permian extinction event 177
Leakey, Louis, Mary and Richard
152
lemurs 168
Lepidodendrons 172
Libellulium longialata 162
limestone 8, 10, 98, 113, 124, 137,
190, 211, 218, 219
dating 189
formation 73
'marble' 204, 205
nummulitic 193
oolitic 73
Solenhofen 13-14
Limulus 13
lizards 101, 167, 178
lobsters 162, 177
lycopods 177, 193
bark 172
roots 172
Lyell, Sir Charles 24, 31

Maladioidella 18
mammals 13, 177
endemism 40-1
mammoths 14
Mantell, Gideon 22
maps, geological 214
Marella splendens 165
marine algae 197
marine deposits see deep sea
deposits
marine depth stratification 90, 91
marine organisms
through geological time 30, 146
and plate tectonics 41-2
marine sedimentation 69, 71-5
marine transgressions 42-4
Marrella 12
marsupials 171
mastodons 171
Megaladapis edwardsi 168
Megatherium 27, 41
Mengeaphis glandulosa 163
Mesolimulus ornatus 161

Mesozoic era 30, 31, 87, 160, 191
meteorites 121
  and early life 120-2
  impacts 118, 120, 181, 183-4
Micraster 149
  M. coranguinum 150
  M. corbovis 150
  M. cortestudinarium 150
microfossils 121
  see also foraminifera
  and radiolaria
micropalaeontology 33, 188, 194
migration, human 157, 158
millipedes 163
mineral deposits 203
Miocene epoch 30, 31, 148
  fossils 161, 162, 170, 171, 174,
  176
Mistaken Point, Canada 133
modern man 154, 157-8
monitor lizards 101
monoplacophora 138, 141
moths 182, 183
mountain ranges 69, 70
mudstones 199, 202, 216

nautiloids 30, 160, 177, 205
Nautilus 68
Neanderthal man 143
  artefacts 158
  skulls 154, 155, 157
Nemagraptus gracilis 32
Neuropteris 199
niches, ecological 90-1, 127
Nummulites 193, 205
  N. gizehensis 193

oceanic deposits/fluctuations
  see deep sea deposits and
  marine transgressions
Ockley brick pit, UK, 7
oil 199-201
oil exploration 33, 200, 202-3
oil industry
  palaeontology 188, 194
oldest fossils 123
Olenoides 12
Oligocene epoch 30, 175
oolitic limestone 73

opal 10
ophiolites 84, 85
Orbulina 148
Ordovician period 29, 30, 42
  corals 80
  extinctions 160
  fossils 10, 18, 32, 34, 51, 60, 63,
  84, 96, 98, 113, 136, 139, 149,
  189
  glacial activity 45
  plankton 77
  rocks 73, 205
ornamental stones 204-5
Orrorin tugensis 152
Orthograptus calcaratus 86
ostracods 190-1, 193
'Out of Africa' theory 158
oysters (Devil's toenails) 54

Palaeocoma egertoni 61
palaeontology see
  micropalaeontology
Paleocene epoch 30
Paleozoic era 30, 31, 39, 82, 85,
  87, 103, 111
palynology 196
Pangaea 38, 39-41, 42, 83
  glaciation 46
Paranthropus 152
  P. robustus 153
Peripatus 115
Permian period 29, 30, 81
  extinctions 81, 177
  fossils 80, 160
  glacial activity 46
  rocks 28
Phacops rana 64
photosynthesis, algal/bacterial
  127, 129
Phyllograptus 105
Piltdown Man 151-2
Pipe Rock, UK 17-18
Placoparia 10
plankton
  algal 85, 132, 198, 200
  global distribution 77
plants
  and continental drift 39
  flowers 174

foliage 173-6, 199
fossil fragments 22
fossilization 10
  microscopic remains 195-7
  roots and bark 172
  sexual differentiation 129
  wood 173
Platanaster ordovicus 60
plate tectonics 37-8, 41
Plegiocidaris coronata 62
Pleistocene epoch 30, 31, 36
  climate 46, 76, 155
  fossils 162, 168
plesiosaurs 177, 178
Pliocene epoch 30, 54
pollen, fossil 195, 196
polychaete worms 53, 91
Populus latior 176
Porana oeningensis 174
prawns 164
Precambrian eon 30, 31, 42
  extinction events 160
  fossils 123-34, 141, 188
  glacial activity 45
  rocks 28, 84, 127, 130, 200
Prediscosphaera cretacea 88
Primaevilium amoenum 123
Pristigenys substriatus 166
prokaryotes 123
Promicroceras 187
  P. marstonensis 204
  P. planicosta 57
Prorichthofenia permiana 80
Proterozoic era 31, 130
Pseudocrinites magnificus 60
pterosaurs 178

quartz, 'shocked' 181
Quaternary period 30

radiolaria 74, 85, 87, 89, 105, 139
radiometric dating of rocks 34-6,
  46
Rana sp. 170
Recent epoch 17, 30
reconstructions, fossil 94-5
reefs 44, 78-82, 201
reptiles 13, 167, 177
Rhus stellariaefolia 175

Rhynie Chert 10
ripple marks 72
rivers
    deltas and estuaries 71
rocks
    angular unconformities 29
    bedding planes 145, 190, 211
    cleavage 21, 211
    dating
        radiometric 34-6, 46
        zonal fossils 32-4, 189,
            191, 193, 194, 195, 197
    folding 20, 211
Rotularia bognoriensis 53
rudists 81, 205
Rusophycus 18

Sabatinca perveta 183
salt domes 201, 202
salt pans 70
sandstone 8, 201, 202, 211, 212,
    218
sauropods 17, 23, 99-103
Scabrotrigonia thoracica 54
Scleractinia 177
sea level changes 43, 44
    and animal evolution 140
    and mass extinctions 48
sea lilies see crinoids
sea spiders 13
sea urchins 61, 62, 78
    evolutionary changes 149-50
    fossil storage 223
    as zonal fossils 30
sedimentary facies 68-78
sedimentary plains 69, 70
sexual differentiation,
    algal/plant 129
shales 8, 20, 21, 84, 105, 106,
    190, 199, 202, 216
    Aberystwyth 67
    Burgess 11-12, 113
    Hunsrück 12
sharks 86-7
    teeth 74, 166
Sigillaria laevigata 172
silica 10, 122, 128, 139, 219
Silurian period 29, 30, 34
    climate 48

fossils 8, 34, 52, 60, 63, 64,
    84, 87, 104, 106, 189
    rocks 28, 67
skeletons, phosphatic 139, 140
skulls, hominid 151-2, 154
slate 21
sloths, giant 27, 41
snakes 101, 178
Solnhofen limestones, Bavaria
    13-14
Spenopteris laurenti 175
spiders 15
Spirifer striatus 52
sponges 80, 139
    glass 49, 85
spores 195, 196, 199
starfish 60
Stigmaria ficoides 172
storage of fossils 222-3
stratigraphy 33
    use of fossils in 32, 91, 188,
        189, 191, 193, 194, 195, 197
Stringer, Chris 157
stromatolites 124, 125, 126, 127
stromatoporoids 80
sundial shells 54

teeth 94, 74, 166
Tertiary period 30, 31, 34, 82,
    89, 148, 160
Tetragraptus approximatus 77
Thalassina anomala 162
Thamnopora cervicornis 50
tillite 45
Timoroblastus coronatus 178
tools 210, 216
trace fossils (ichnofossils) 17-19,
    133-4, 211
tracks
    dinosaurs 17, 99, 100
    trilobites 18
Treatise on Invertebrate
    Palaeontology 220-1
Triassic period 29, 30, 31, 160
    fossils 59, 173
    rocks 205
Tribrachidium 132, 133
Triceratops 101
trilobites 10, 12, 22, 63, 64, 135,

137, 203, 206
    agnostid 84, 85
    extinctions 160, 177
    eyes and vision 96-7
    life habits 95-9
    phacopid 13
    tracks 18
    as zonal fossils 30, 33, 34
Trinucleus fimbriatus 63
turbidites 75-6, 83-4, 85, 211
Tylocidaris clavigera 62
Typhis pungens 55
Tyrannosaurus rex 207

unconformities 29

Vendian (Ediacarian) era
    fossils 130, 132-3
vertebrates, terrestrial
    Cretaceous survival 179
viral epidemic theories 122
Virgatosphinctes 40
Visbyshaera oligofurcata 87
volcanic activity 38, 39, 75, 79,
    118, 120, 183
    and mass extinctions 48
Voluta muricina 55

Wadi Kharaza, Egypt 125
Wegener, Alfred 46
whales 74
wood, fossil 206
worms 12, 17
    annelid 132
    polychaete 53, 91
    velvet 115

zoning and zonal fossils 32-4,
    189, 191, 193, 194, 195, 197
Zygolophodon atticus 171